# Sensors and Control Systems
# in Arc Welding

# Sensors and Control Systems
# in Arc Welding

*Edited by*

## Hirokazu Nomura

*Chairman, Technical Commission on Welding Processes*
*Japan Welding Society,*
*Visiting Professor, Mie University, and Director.*
*Engineering Research Center*
*NKK Corporation, Kawasaki, Japan*

*Technical editor*

## John E. Middle

*Department of Manufacturing Engineering,*
*Loughborough University of Technology, UK*

*With foreword by*

## Isao Masumoto

**CHAPMAN & HALL**
London · Glasgow · New York · Tokyo · Melbourne · Madras

**Published by Chapman & Hall, 2–6 Boundary Row, London SE1 8HN, UK**

Chapman & Hall, 2–6 Boundary Row, London SE1 8HN, UK

Blackie Academic & Professional, Wester Cleddens Road, Bishopbriggs, Glasgow G64 2NZ, UK

Chapman & Hall Inc., One Penn Plaza, 41st Floor, New York NY10119, USA

Chapman & Hall Japan, Thomson Publishing Japan, Hirakawacho Nemoto Building, 6F, 1-7-11 Hirakawa-cho, Chiyoda-ku, Tokyo 102, Japan

Chapman & Hall Australia, Thomas Nelson Australia, 102 Dodds Street, South Melbourne, Victoria 3205, Australia

Chapman & Hall India, R. Seshardi, 32 Second Main Road, CIT East, Madras 600 035, India

English language edition 1994

© 1994 Japan Welding Society

Original Japanese language edition © 1991 Technical Commission on Welding Processes, Japan Welding Society.

Typeset in 10/12 pt Times by Interprint Limited, Malta.
Printed in England by Clays Ltd, St Ives plc

ISBN 0 412 47490 5

∞ Printed on permanent acid-free text paper, manufactured in accordance with the proposed ANSI/NISO Z 39.48–199X and ANSI Z 39.48–1984

# Contents

Contributors     ix
**Members of the Editorial Committee**     xiii
**Foreword**     xv
**Preface**     xvii

**SECTION ONE General Review**

1. Introduction to sensing systems and their application in arc
   welding processes     1
   1.1 Role of sensors     1
   1.2 Definition and classification of sensors     2
   1.3 Definition and classification of control systems     7

2. State-of-the art review of sensors and their application in arc
   welding processes     18
   2.1 Contact sensors     18
   2.2 Temperature sensors     22
   2.3 Arc phenomena sensors (through-the-arc sensors)     25
   2.4 Optical sensors     32
   2.5 Electromagnetic sensors     39
   2.6 Sound sensors     40

3. Control systems     44
   3.1 Seam tracking and deposition control     44
   3.2 Welding process control     45
   3.3 Seam tracking based on fuzzy control     48

4. Basis for welding process control     53
   4.1 Modern control theory     53
   4.2 Molten pool width control     56
   4.3 Adaptive control     58
   4.4 Fuzzy control     64

5. Sensor techniques and welding robots     69
   5.1 Present status of robot sensors     70
   5.2 Connection between sensor and robot     72
   5.3 Robot off-line programming     73

6. Analysis of questionnaire results on sensor applications to
   welding processes                                              76

7. Future trends                                                  86

**SECTION TWO Applications of Sensors to the Welding Process**

8. Laser sensing methods for groove tracking control             88
   *Hirohisa Fujiyama*

9. An automatic multilayer welding system with laser sensing      95
   *Gouhei Iijima, Sumihiro Ueda, Masaaki Hirayama, Kaoru
   Kawabe and Osamu Takamiya*

10. On-line visual sensor system for arc-welding robots          104
    *Makoto Endo*

11. Welding robot with visual seam tracking sensor               111
    *Akira Hirai, Nobuo Shibata, Toshio Akatsu and Kyoichi
    Kawasaki*

12. An arc-welding robot with a compact visual sensor            118
    *Takao Bamba*

13. A robot with visual sensor for three-dimensional multilayer
    welding                                                      127
    *Teppei Asuka, Jyunzou Komatu, Kouichi Satou and Michio
    Kitamura*

14. A visual arc sensor system                                   132
    *Shinji Okumura and Seigo Nishikawa*

15. Fuzzy control of $CO_2$ short-arc welding                    137
    *Kenji Ohshima, Satoshi Yamane, Haruhiko Iida, Shunjiang
    Xiang, Yasuchika Mori and Takefumi Kubota*

16. Application of a fuzzy neural network to welding line tracking  147
    *Kenji Ohshima, Satoshi Yamane, Yasuchika Mori and Peijun Ma*

17. Magnetic control of the arc in high-speed TIG welding        154
    *Shizuo Ukita, Taisuke Akamatsu and Kan'ichiro Shimizu*

18. Adaptive control of welding conditions using visual sensing  160
    *Ken Fujita and Takashi Ishide*

19. Automatic welding system with laser optical sensor for heavy-
    walled structures                                            168
    *Hirokazu Wada, Yukio Manabe and Shigeo Inoue*

20. Group-control system for narrow-gap MIG welding              176
    *Hirokazu Wada, Yukio Manabe, Yoshinori Hiromoto and Hideki
    Yoritaka*

21. Automatic control technique for narrow-gap GMA welding      182
    *Hiroshi Watanabe, Yoshihide Kondo and Katsunori Inoue*

22. An automatic control system for one-sided submerged arc
    welding with flux copper backing      191
    *Nobuyuki Okui*

23. Touch sensor and arc sensor for arc-welding robots      199
    *Jun Nakajima, Takeshi Araya, Shinnichi Sarugaku, Kooji Iguchi
    and Yuuji Takabuchi*

24. Application of arc sensors to robotic seam tracking      209
    *Shunji Iwaki*

25. Dynamic analysis of arc length and its application to arc
    sensing      216
    *Etsuzou Murakami, Katsuya Kugai and Hideyuki Yamamoto*

26. Robot welding with arc sensing      228
    *Hiroshi Fujimura, Eizo Ide and Hironori Inoue*

27. Arc sensing using fuzzy control      238
    *Hiroshi Fujimura, Eizo Ide and Shuta Murakami*

28. Development and application of arc sensor control with a
    high-speed rotating-arc process      247
    *Hirokazu Nomura, Yuji Sugitani, Yukio Kobayashi and
    Masatoshi Murayama*

29. Groove tracking control by arc-welding current      257
    *Mitsuaki Otoguro*

30. Automatic seam tracking and bead height control by arc
    sensor      266
    *Hirokazu Nomura, Yuji Sugitani and Naohiro Tamaoki*

31. Through-the-arc sensing control of welding speed for one-sided
    welding      279
    *Yasuyoshi Kitazawa*

32. Some types of wire ground sensors      285
    *Naoki Takeuchi*

33. Application of a touch sensor to an arc welding robot      292
    *Hisahiro Fukuoka*

34. Automatic welding for LNG corrugated membranes      300
    *Hirokazu Nomura and Tadashi Fujioka*

35. Welding process monitoring by arc sound      308
    *Masami Futamata*

# Contents

36. A vibrating reed sensor for tracing the centre of narrow grooves    316
*Katsuyoshi Hori, Yoshihide Kondo, Kazuki Kusano and Hiroyasu Enomoto*

37. Development of an ultra-heat-resistant electromagnetic sensing system for automatic tracking of welding joints    323
*Kazuhiko Wakamatsu and Yasuo Kondo*

38. Motion generation in an off-line programming system for an arc-welding robot    333
*Hitoshi Maekawa*

39. Application of an off-line teaching system for arc-welding robots    342
*Teruyoshi Sekino and Tsudoi Murakami*

40. Development and availability of an IC card welding system for automatic welding    350
*Kazuhiro Takenaka, Yoshio Imajima and Susumu Ito*

41. Development of a remote-controlled circumferential TIG welding system    360
*Tomoya Fujimoto and Atsushi Shiga*

42. In-process control systems in one-side SAW    369
*Hirokazu Nomura, Yukihiko Satoh and Yoshikazu Satoh*

43. Application of an arc sensor and visual sensor on arc-welding robots    382
*Teruyoshi Sekino*

44. An intelligent arc welding robot with simultaneous control of penetration depth and bead height    390
*Yuji Sugitani, Yasuhiko Nishi and Toyoyuki Satoh*

45. Au NC welding robot for large structures    400
*Takaaki Ogasawara*

46. Development of a fully automatic pipe welding system    408
*Akio Tejima*

**Index**    417

# Contributors

**Taisuke Akamatsu**, Department of Mechanical Engineering, Kogakuin University.

**Toshio Akatsu**, Mechanical Engineering Research Laboratory, Hitachi Ltd.

**Takeshi Araya**, Mechanical Engineering Research Laboratory, Hitachi Ltd.

**Teppei Asuka**, Hitachi Works, Hitachi Ltd.

**Takao Bamba**, Industrial Electronics & Systems Development Laboratory, Mitsubishi Electric Corporation.

**Makoto Endo**, Mechatronics Products Department, Ishikawajima-Harima Heavy Industries Co. Ltd.

**Tomoya Fujimoto**, Engineering Division, Kawasaki Steel Corporation.

**Hiroyasu Enomoto**, Yokahama Works, Babcock-Hitachi KK.

**Hiroshi Fujimura**, Nagoya Research & Development Centre, Mitsubishi Heavy Industries Ltd.

**Tadashi Fujioka**, Engineering Research Centre, NKK Corporation.

**Ken Fujita**, Takasago Research & Development Centre, Mitsubishi Heavy Industries Ltd.

**Hirohisa Fujiyama**, Machinery Department, Nippon Steel Welding Products & Engineering Co. Ltd.

**Hisahiro Fukuoka**, Industrial Machinery Division, Shin Meiwa Industry Co. Ltd.

**Masami Futamata**, Faculty of Engineering, Kitami Institute of Technology.

**Akira Hirai**, Mechanical Engineering Research Laboratory, Hitachi Ltd.

**Masaaki Hirayama**, Kawasaki Heavy Industries Ltd.

**Katsuyoshi Hori**, Kure Research Laboratory, Babcock-Hitachi KK.

**Yoshinori Hiromoto**, Hiroshima Research & Development Centre, Mitsubishi Heavy Industries Ltd.

**Eizo Ide**, Nagasaki Research & Development Centre, Mitsubishi Heavy Industries Ltd.

**Kooji Iguchi**, Narashino Works, Hitachi Ltd.

**Haruhiko Iida.**

**Gouhei Iijima**, Kawasaki Heavy Industries Ltd.

**Yoshio Imajima**, Keihin Product Operations, Toshiba Co.

**Hironori Inoue**, Nagasaki Research & Development Centre, MHI Ltd.

**Katsunori Inoue**, Welding Research Institute, Osaka University.

**Shigeo Inoue**, Hiroshima Research & Development Centre, Mitsubishi Heavy Industries Ltd.

**Takashi Ishide**, Takasago Research & Development Centre, Mitsubishi Heavy Industries Ltd.

**Susumu Ito**, Keihin Product Operations, Toshiba Co.

**Shunji Iwaki**, Industrial Machinery Division, Shin Meiwa Industry Co. Ltd.

**Kaoru Kawabe**, Kawasaki Heavy Industries Ltd.

**Kyoichi Kawasaki**, Narashino Works, Hitachi Ltd.

**Michio Kitamura**, Hitachi Works, Hitachi Ltd.

**Yasuyoshi Kitazawa**, Welding Division, Kobe Steel Ltd.

**Yukio Kobayashi**, Engineering Research Centre, NKK Corporation.

**Jyunzou Komatu**, Hitachi Works, Hitachi Ltd.

**Yasuo Kondo**, Manufacturing Engineering Department, Mitsubishi Heavy Industries Ltd.

**Yoshihide Kondo**, Kure Works, Babcock-Hitachi KK.

**Takefumi Kubota**, Himeji Institute of Technology.

**Katsuya Kugai**, Mechatronics Products Division, Daihen Corporation.

**Kazuki Kusano**, Kure Works, Babcock-Hitachi KK.

**Munehaju Kutsuna**, Faculty of Engineering, Negoya University.

**Peijun Ma**, Electrical Engineering Department, Saitama University.

**Hotoshi Maekawa**, Faculty of Engineering, Saitama University.

**Yukio Manabe**, Hiroshima Research & Development Centre, Mitsubishi Heavy Industries Ltd.

**Yasuchika Mori**, Electrical Engineering Department, Saitama University.

**Etsuzou Murakami**, Mechatronics Products Division, Daihen Corporation.

**Shuta Murakami**, Qushu Institute of Technology.

**Tsudoi Murakami**, Welding Division, Kobe Steel Ltd.

**Masatoshi Murayama**, Engineering Research Centre, NKK.

**Jun Nakajima**, Industrial Systems and Equipment Group, Hitachi Ltd.

**Yasuhiko Nishi**, Engineering Research Centre, NKK Corporation.

**Seigo Nishikawa**, Application Engineering Section, Yaskawa Electric Mfg Co. Ltd.

**Hirokazu Nomura**, Engineering Research Centre, NKK Corporation.

**Takaaki Ogasawara**, Welding Division, Kobe Steel Ltd.

**Kenji Ohshima**, Electrical Engineering Department, Saitama University.

**Nobuyuki Okui**, Industrial Engineering Department, Ishikawajima-Harima Heavy Industries Co. Ltd.

**Shinji Okumura**, Application Engineering Section, Yaskawa Electric Mfg Co. Ltd.

**Mitsuaki Otoguro**, Research Institute, Nippon Steel Welding Products and Engineering Co. Ltd.

**Shinnichi Sarugaku**, Hitachi Keiyou Engineering Co. Ltd.

**Toyoyuki Satoh**, Keihin Factory, Nippon Sanso KK.

**Yoshikazu Satoh**, Engineering Research Centre, NKK Corporation.

**Yukihiko Satoh**, Pipeline Projects, NKK Corporation.

**Kouichi Satou**, Hitachi Works, Hitachi Ltd.

**Teruyoshi Sekino**, Welding Division, Kobe Steel Ltd.

**Nobuo Shibata**, Mechanical Engineering Research Laboratory, Hitachi Ltd.

**Atsushi Shiga**, Engineering Division, Kawasaki Steel Corporation.

**Kan'ichiro Shimizu**, Department of Mechanical Engineering, Kogakuin University.

**Yuji Sugitani**, Engineering Research Centre, NKK Corporation.

**Yuuji Takabuchi**, Narashino Works, Hitachi Ltd.

**Osamu Takamiya**, Kawasaki Heavy Industries Ltd.

**Naoki Takeuchi**, Welding Division, Kobe Steel Ltd.

**Naohiro Tamaoki**, Engineering Research Centre, NKK Corporation.

**Kazuhiro Takenaka**, Keihin Product Operations, Toshiba Co.

**Akio Tejima**, Industrial Engineering Department, Ishikawajima-Harima Heavy Industries Co. Ltd.

**Sumihiro Ueda**, Kawasaki Heavy Industries Ltd.

**Shizuo Ukita**, Department of Mechanical Engineering, Kogakuin University.

**Masao Ushio**, Welding Research Institute, Osaka University.

**Hirokazu Wada**, Hiroshima Research & Development Centre, Mitsubishi Heavy Industries Ltd.

**Kazuhiko Wakamatsu**, Shinryo High Technology and Control Co. Ltd.

**Hiroshi Watanabe**, Yokohama Works, Babcock-Hitachi KK.

**Shujiang Xiang.**

**Hideyuki Yamamoto**, Welding Products Division, Daihen Corporation.

**Satoshi Yamane**, Maizuru College of Technology.

**Hideki Yoritaka**, Hiroshima Research & Development Centre, Mitsubishi Heavy Industries Ltd.

# Members of
# the Editorial Committee

# Foreword

The Research Commission on 'Welding Processes' in the Japanese Welding Society (R.C.W.P., JWS) has compiled two state-of-the-art reports in Japan: *One Sided Automatic Welding Processes*, presented at the IIW annual assembly in Kyoto in 1969; and *Narrow Gap Welding*, presented at the IIW annual assembly in Tokyo in 1986. These reports were well received by assembly participants and have subsequently triggered many active discussions within the IIW Commission XII, the R.C.W.P. and the Commission 12 of the Japan Institute of Welding (JIW 12).

One of the most important aims of the IIW's activities has been to produce an up-to-date report, incorporating the proceedings described above, for the international welding industry. To this end, the Chairman of the R.C.W.P., Dr H. Nomura, has recently completed this guide book *Sensors and Control Systems in Arc Welding*. The project has required the co-operation of many researchers and engineers in Japan including Professor Ushio from the University of Osaka and Mr Yamamoto from the OTC Co. who are also acting as co-ordinators of special projects for the IIW Sub-commissions XII-A and XII-C respectively.

I am sure that *Sensors and Control Systems in Arc Welding*, as a report of one of the most important projects currently undertaken by the IIW Commission XII, will receive high international acclaim and I would like to express my greatest respect and gratitude to those who have co-operated in its publication. I am particularly grateful to Dr Nomura for his leadership as the R.C.W.P. Chairman and to Professor Matsunawa from the Osaka University for his contribution as Editor-in-chief.

Professor Dr Isao Masumoto
Vice-chairman of IIW Commission XII
Chairman of IIW Sub-commission XII-A

# Preface

The Japan Welding Society's Technical Commission on Welding Processes has now held 131 meetings since it was first established. Today, the worldwide trend in industrial products is towards ever smaller and lighter products. Despite this, the Commission is still as active as ever, in response to the growing demand for automatic welding and labour-saving processes. The importance of the basic technologies of welding, sensing and control has been clearly recognized.

In 1984, the Commission embarked on the publication of a series of books that would provide a guide to the state of the art of Japanese welding technology. Compiled under the guidance of Dr Itsuhiko Sejima and Professor Fukuhisa Matsuda, respectively Chairman and Vice-Chairman at the time, the first book in the series dealt with the narrow-gap welding process. It served as a valuable reference book, not just for the members of the Commission, but for the members of the Japan Welding Society as a whole. It was subsequently translated into English to reach an even wider audience.

This volume is the second edition of a similar guide to the state-of-the-art welding technologies which lay the foundation for sensors, sensing systems, measurement and control. Its publication is most timely. It has been completed thanks to the cooperation of senior personnel in universities, independent research institutes, manufacturers and users, coordinated by Professor Akira Matsunawa, Vice-Chairman of the Commission. I wish to express my sincere gratitude to everyone who has helped towards the publication of this book. I have no doubt that it will be well received, and will be a valuable source of information to all the members of the Commission, to the Japan Welding Society, and to industry worldwide.

Dr Hirokazu Nomura
Chairman, Technical Commission on Welding Processes
Japan Welding Society

# Introduction to sensing systems and their application in arc welding processes

## 1.1 ROLE OF SENSORS

The last decade has seen dramatic technical breakthroughs in automatic welding equipment: arc welding robots, for example. A survey by the Japan Industrial Robot Association revealed that about 6000 arc welding robots were manufactured in 1988, nearly 50% of which were equipped with sensors.

However, a universal sensor has not yet been developed. The current state of the art in the robot industry requires suitable robots and sensors to be selected for specific purposes. Sensor development is therefore essential for progress in automation or improvement of the artificial intelligence of the welding operation [2–4].

Disturbances or variability related to the objects to be welded when 'viewed' from the arc welding robot include:

1.  work shape error;
2.  setting error;
3.  variation in groove shapes;
4.  tack beads;
5.  welding deformation;
6.  jig error.

Disturbances related to the welding process include:

1.  arc light;
2.  arc heat;
3.  spatter;
4.  electromagnetic field;
5.  bending and deflection of electrode wire;

*Sensors and Control Systems in Arc Welding.* Edited by Hirokazu Nomura.
Published in 1994 by Chapman & Hall, London. ISBN 0 412 47490 5

6. fluctuation in wire feeding speed;
7. wear of contact tips;
8. change in welding conditions and arc shapes.

If it was possible to eliminate many of these disturbances during the welding operation, more favourable results could be expected. However, in practice it is virtually impossible to remove all these disturbances directly. Therefore, they must be indirectly eliminated by the use of sensors, which measure the disturbances and provide control imformation to allow compensations in process parameters and manipulations.

The conditions that the sensors must fulfil include:

1. the capability to maintain the accuracy proper to a specified welding process;
2. freedom from the influence of welding process-induced disturbances (such as light, heat, fume, spatter and electromagnetics);
3. satisfactory durability;
4. low cost;
5. easy maintenance;
6. compact size and light weight;
7. wide range applicability.


## 1.2   DEFINITION AND CLASSIFICATION OF SENSORS

Sensors for arc welding were defined in a 1982 report [1] as follows:

A detector, if it is capable of monitoring and controlling welding operation based on its own capacity to detect external and internal situations affecting welding results and transmit a detected value as a detection signal, is called as a sensor. Moreover, its whole control device is defined as a sensor system (control system).

The external situations referred to here cover the presence of component obstacles, changes in welding groove dimensions and welding line positions, and the presence of tack welds. The internal situations cover such factors as welding arc shape, the dimensions of the molten pool, penetration state, temperature distribution at welds, and arc sound pressure, which are all related to welding phenomena themselves.

However, the following are excluded from the objectives of the arc welding sensors:

- sensors used to control the equipment, such as welding power supply and wire feeding devices;
- sensors used to control the equipment, such as welding robots and manipulators;
- sensors used to record and inspect welds for quality control.

Table 1.1 shows the detection objectives of various types of sensor and examples of their applications: both those that fall within the scope of this book and those that fall outside it. Table 1.2 shows the classification of sensors by principles based on each component.

**Table 1.1** Functional range of sensors used in arc welding

| Detection objective | Examples |
|---|---|
| **Included** | |
| Components | Contact or non-contact sensors that can recognize welding position, groove shape, and obstructions |
| Welding characteristics | Sensors that can recognize arc length, wire extension length, arc shape, weld pool dimensions, external appearance of bead, penetration conditions, and arc sound |
| **Excluded** | |
| Welding machines | Sensors for detecting quantity of shield gas flow (pressure), coolant water pressure, current overload and wire feed torque |
| Automatic welding machines | Sensors that control position by encoders or potentiometers and/or sensors that control speed by tachometers and accelerometers |
| Welding quality control | Sensors for inspecting the results of welding by X-ray and ultrasound and sensors for recording welding parameters |

**Table 1.2** Classifications of sensors

| Sensor type | Units in the sensor configurations |
|---|---|
| **Contact** | |
| Contact probes | Microswitches, potentiometers and differential transformers (DTF) |
| Electrode contact | Voltage and current for contact detection that is applied to the W electrode or the welding wire |
| Temperature | Thermocouples and thermistors |
| **Non-contact** | |
| Temperature | Photothermometers and infrared thermometers |
| Arc phenomena | Welding current arc voltage, wire feed speed, number of shorts, number of peak current anomalies |
| Optics | Point sensors (phototransistors ad photodiodes), linear sensors (CCD, MOS and PSD), and area (image) sensors (CCD, MOS, PSD and ITV)* |
| Sound | Variable sound pressure or ultrasonic sound pressure detector probes |

*CCD, charge-coupled device; MOS, metal oxide semiconductor; PSD, position-sensitive detector; ITV, industrial television.

### 1.2.1   Contact probe

A contact probe is a sensor which outputs distance changes as electrical signals from a tactile probe applied to a groove. The probe can either have one degree of freedom (Figure 1.1(a)), or two degrees of freedom (Figure 1.1(b)). There are two other types of sensors: those which output an analogue signal proportional to distance using a potentiometer or a differential transformer, and those which output an on/off signal based on the application of a limit switch.

These sensors have been widely used for a long time, and are simple to handle and maintain.

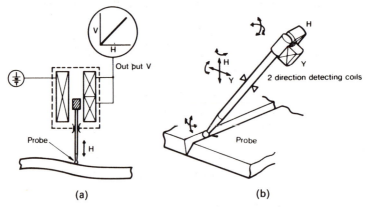

(a)                                    (b)

**Fig. 1.1** Contact probe sensors: (a) differential transformer with one degree of freedom; (b) contact sensor with two degrees of freedom.

### 1.2.2   Electrode contact sensor

This type of sensor detects the change in electric current or voltage flowing when a welding wire is placed into contact with a base material (Figure 1.2), and obtains the coordinates of a contact point from the coordinates of a robot or an automatic machine in service. It is generally called a *wire touch*, *wire earth* or *touch* sensor. A voltage ranging from 300 to 600 V and from 50 to 600 Hz is applied to obtain steady contact. The wire extension may be controlled by automatically cutting it to length prior to sensing.

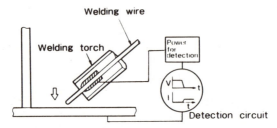

**Fig. 1.2** Principles of electrode contact sensors.

Alternatively, the wire extension can be compared with a reference point and revised based on the comparison. A wire lock mechanism is installed to improve the accuracy of the coordinates.

### 1.2.3 Temperature sensor

This sensor is used for adaptive control of welding conditions to optimize welding bead shape or uranami (penetration bead shape). A contact thermometer, such as a thermocouple or thermistor, or a non-contact type, such as a light thermometer or an infrared thermometer, is used to measure the temperature of material in or nearer the weld. Figure 1.3 shows the standard application temperature range of practical thermometers.

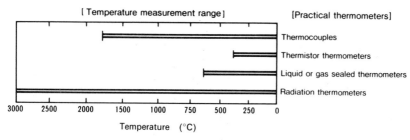

**Fig. 1.3** Temperature measurement range of practical thermometers.

### 1.2.4 Arc phenomena sensor

This is a sensor which utilizes the changes in welding current induced by the change in torch height in arc welding, as illustrated in Figure 1.4. Generally, it is called an *arc sensor*, *ACC* (*automatic current control*) *sensor*, *AVC* (*automatic voltage control*) *sensor* or *through-the-arc sensor*. It allows the application of arc voltage as well since the output characteristic of the power supply is slightly drooping (2–3 V/100 A). In TIG welding, arc voltage is used since the power supply has the drooping characteristics.

There are some other examples where the height of a torch $H = f (I, V, v)$ is calculated based on detecting the welding current $I$, arc voltage $V$ and wire feeding speed $v$.

Since the welding current or arc voltage may fluctuate dramatically, various approaches are used:

1. smoothing the welding current or arc voltage with a filter;
2. applying a specific frequency component based on the application of Fourier trasformation;
3. picking up information only in the arc period.

Adaptive control of welding conditions is effected by detecting such factors as the number of short circuits of the wire and the number of abnormal current peaks.

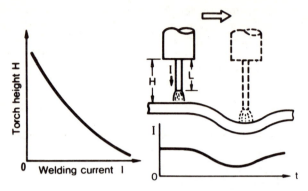

**Fig. 1.4** Principles of arc sensors.

### 1.2.5 Electromagnetic sensor

This sensor is designed to provide an analogue output of the height $H$ relative to the base material by exciting coils fastened round a core with a.c. input and connecting differentially two detection coils lapped as illustrated in Figure 1.5. This enhanced design, with differential connection and lap winding, is introduced to prevent the generation of welding-induced disturbances of magnetic field and temperature.

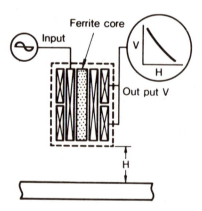

**Fig. 1.5** Electromagnetic sensor.

### 1.2.6 Optical sensor

Various types of optical sensors designed to detect visual light or laser light are used, but they are largely classified into point sensors, linear sensors, and area (image) sensors; based on the shape of the light pattern received.

### (a)  Point sensor

The point sensor, which uses a photodiode or a phototransistor, is designed to detect a point light source reflected or emitted from a base material based on on/off or analogue quantities.

### (b)  Linear sensor

The linear sensor is designed to detect a line of light reflected or emitted from a base material in one dimension. Its application covers:

1. several point sensors laid out in a linear array;
2. CCD and MOS used as a digital component;
3. PSD used as an analogue component.

### (c)  Area (image) sensor

The area (image) sensor is designed to detect the light reflected or emitted from a base material in two dimensions. The application covers:

1. ITV;
2. CCD and MOS where the digital components are laid out in two dimensions;
3. PSD used as an analogue component.

### 1.2.7  Sound sensor

The sound sensor is classified into types based on microphones, to detect audible sound during arc welding, and types based on ultrasonic sensors designed to transmit and receive ultrasonic waves. There are two types of ultrasonic sensor. One is integrated into one piece, but provides independent transmitting and receiving functions, while the other type is used on a time-sharing system.

## 1.3  DEFINITION AND CLASSIFICATION OF CONTROL SYSTEMS

A *control system* constitutes an attempt to control a welding process based on the information obtained from a sensor. The control system may be classified by application:

1. seam tracking control of a welding line;
2. adaptive control for welding conditions;
3. seam tracking control and adaptive control;
4. welding monitoring.

### 1.3.1  Seam tracking control

This is the most popular form of control system. It is designed to obtain three-dimensional or two-dimensional information directly or indirectly from a sensor and hence control the tip locus and angle of a welding torch attached to a welding robot or an automatic welding machine.

One or several sensors are used to control the welding path and its start and end point. When an attempt is made to weld a thick plate with a multipass welding method, seam tracking control may be adopted to detect the position of each welding pass while a shift function may be adopted to trace the locus of the previous pass while deviating its locus by some distance.

#### (a)  Seam tracking control with contact probe

This control system is designed to follow the welding path by fixing a probe to a welding torch (Figure 1.1) so that they may move as an entity. Several sensors with one degree of freedom may be used to control torch angle or measure groove width.

#### (b)  Seam tracking control with electrode contact sensor

This control system is designed to bring the welding wire into contact with the surfaces of the component at several points as illustrated in Figure 1.6. It obtains the coordinates at the intersection point P by arithmetic calculation. The correct position of the welding torch is obtained from a corresponding offset to the proprogrammed point on the robot path. Such a method may be used to find the start of the weld line, using some other sensor to follow the weld seam, or may be used to find both ends of the weld line, permitting relocation of the whole robot path.

#### (c)  Seam-tracking control based on arc phenomena

This seam-tracking control system is based on the application of arc phenomena and comprises a combination of various elements and factors as shown in Figure 1.7.

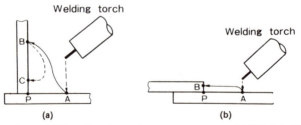

**Fig. 1.6** Detecting weld lines by electrode contact sensors; (a) T-joint; (b) lap joint.

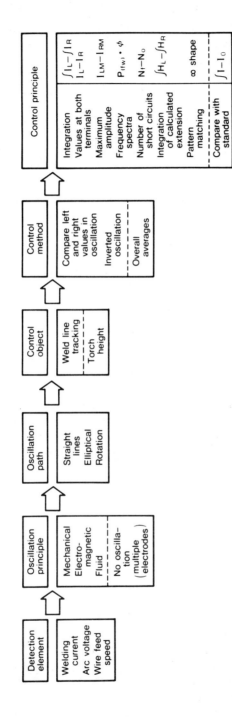

**Fig. 1.7** Tracking control systems categorized by arc sensor.

In this control system, when the welding torch is oscillated in a groove (Figure 1.8), the welding current $I$ changes as illustrated. To comply with this, the height $H$ of the torch is controlled; the welding line may also be traced by processing the electric values on both sides of the oscillation. The detection elements comprise various kinds in addition to the welding current shown in Figure 1.7; in fact, many different control systems and principles have been developed.

Weaving actions force the welding torch to oscillate mechanically. Other methods of control deflect the arc with electromagnetic force without oscillating the torch, or bend the arc with a thermal pinch effect by passing gas near the arc. Oscillation patterns include linear reciprocating motion, circular motion and rotating motion.

Since this control can be adopted only during weld tracking, it is necessary to use a contact probe or an electrode contact sensor to detect the start and end points of welding.

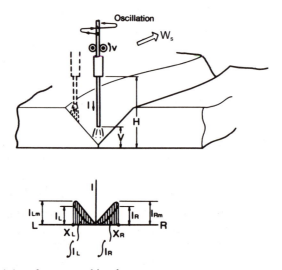

**Fig. 1.8** Principles of seam tracking by arc sensor.

### (d) Seam tracking control with electromagnetic sensor

This control system tracks the welding line by using two electromagnetic sensors arranged at right angles as illustrated in Figure 1.9(a). The distance of each of the two sensors is maintained with respect to the base components, thereby presenting the torch axis at the correct angle to the joint intersection (weld path). Seam tracking control is carried out by detecting the output changes produced when an attempt is made to oscillate the sensor, as illustrated in Figure 1.9(b). There are some examples in which a butt joint may be traced by detecting a magnetic leak quantity induced by welding current as illustrated in Figure 1.9(c) and hence controlling equilibrium on both sides.

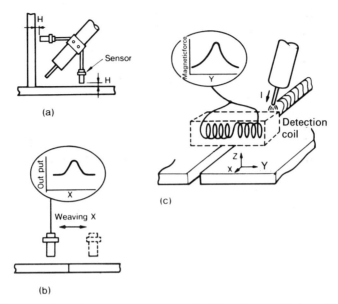

**Fig. 1.9** Seam tracking by magnetic sensors: (a) fillett joints; (b) butt joints; (c) detecting leaked magnetic field.

*(e) Seam tracking control with optical sensor*

Control systems that use optical elements are applied in various situations (Figure 1.10). They are also used to detect and control tacking beads, welding start and end points, and the position of the wire tip, in addition to seam tracking control and torch height control. They are also widely used for adaptive control of welding conditions, which will be described later.

1. *Seam tracking control with non-contact point sensor*: This is a system which carries out seam tracking control by oscillating or rotating a point sensor as illustrated in Figure 1.11(a) to let the sensor cross over a welding line and process the quantity of light received and the position of the sensor at that time in two dimensions.
2. *Seam tracking control with linear sensor*: This is a detection system based on a linear sensor. A point light source, such a semiconductor laser, irradiates the base material as illustrated in Figure 1.11(b). Its light is detected by a linear sensor arranged at a fixed angle $\theta$. The light source and light detector are installed in a single unit. This control system employs the principle that a light-receiving position changes by $\Delta h$ when its distance from the base material changes by $\Delta H$.

   This sensor comprises a system which controls the height of the torch and a system which detects the variation of height of the sensor relative to the base material and hence the shape of a lap joint or groove by oscillating or rotating the sensor across the welding line. Such sensors are also used to track the welding line.

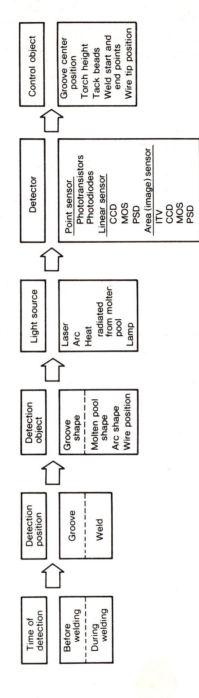

**Fig. 1.10** Seam-tracking control systems categorized by optical sensor.

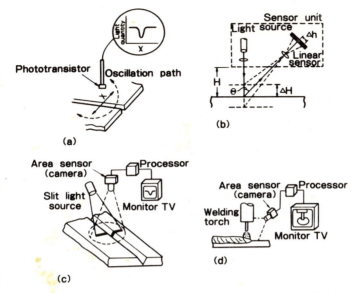

**Fig. 1.11** Seam-tracking control by optical sensor: (a) point sensor; (b) linear sensor; (c) area sensor, type 1; (d) area sensor, type 2.

3. *Seam tracking control with area sensor*: An area sensor comprises a system which receives a stripe of light irradiated to a base material, processes the image, and hence judges the welding position, as illustrated in Figure 1.11(c). An alternative system is shown in Figure 1.11(d). It processes an image obtained by directly viewing the welding region, enabling the position of the groove and welding wire, and the size of welding bead, to be determined.

### (f) Sound sensor

An ultrasonic sensor, as shown in Figure 1.12, is capable of measuring the distance between the sensor and a base material by determining the time *t* from the point at which a pulse is transmitted to the point at which

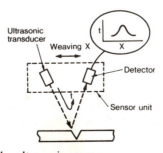

**Fig. 1.12** Seam tracking by ultrasonic sensor.

the reflected pulse is received. It is thus possible to detect the position of a groove by oscillating this unit and hence to trace the welding line. Alternatively a single sensor can be used, which is capable of both transmitting and receiving pulses with the same unit.

### (g)   *Seam-tracking control with a combination of sensors*

A welding robot seam-tracking control system may adopt a variety of sensors which combine functions and detect welding start and end points, and the welding line both prior to and during welding operation. Typical examples of combined sensors are:

1. a wire touch sensor and an arc sensor;
2. a wire touch sensor, an arc sensor and an electromagnetic sensor;
3. an arc sensor and an electromagnetic sensor.

### 1.3.2   Adaptive control of welding conditions

Adaptive control is a system which controls weld quality in real time based on the information obtained from sensors. Recently, various research activities and development results have been reported in this area.

The factors controlled include:

1. welding bead shape (height and width);
2. back bead shape;
3. penetration depth;
4. deposit amount;
5. leg length.

When interpreted in a broad sense, the shape of the molten pool or arc states (such as shape or sound) are included in this category.

Control elements comprise welding current, arc voltage, welding speed and arc position. Adaptive control may involve a direct system, which uses a sensor to measure a control object directly, or an indirect system which (for example) measures the surface temperature of a control object so as to control the shape of back beads.

Various kinds of control technology are used for adaptive control: they include proportion integral control, digital control, fuzzy control, modern control theories, and artificial intelligence (AI).

### (a)   *Adaptive control with temperature sensor*

This is a recently developed system which controls welding parameters, such as peak current in pulse arc welding, d.c. arc current and welding speed, by measuring the temperature of the surface of the base material near the weld bead or the temperature, thermal image and luminance distribution of the

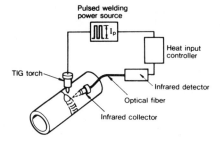

**Fig. 1.13** Principle of penetration control.

molten pool in order to control the penetration depth (width of back bead) of TIG weld (stainless steel) and heat input. Figure 1.13 shows an example of a system which controls the shape of back beads of stainless steel pipe, using an infrared thermometer.

### (b)    Adaptive control with arc phenomena sensor

One example is a system developed to control the amount of deposit by the application of arc phenomena (through-the-arc sensing). As illustrated in Figure 1.14, this system is capable of tracking a welding groove and detecting groove width by moving a welding torch in the groove under constant extension control and reversing its control movement direction when the height of the torch reaches a specified value. It is also capable of computing a deposit cross-sectional area $A$ from a detected groove width. Therefore, when the groove width changes, the variation $\Delta A$ of the cross-sectional area can be computed. This makes it possible to compute the welding speed $v$ at which a constant bead height will be maintained.

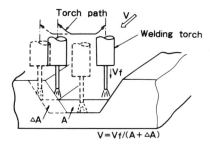

**Fig. 1.14** Method of controlling quantity of deposited metal.

Other examples of such adaptive control systems include:

1.  a system which controls welding speed and the height of back beads for a constant value since the frequency of short-circuiting is associated with the shape of the back beads in the first weld pass;

2. a system which detects turbulence of the welding current waveform and controls the welding speed so that the shape of the back beads may be kept constant;
3. a system which keeps the width of the back beads constant by controlling the welding current with detection of the voltage between a base material and a backing copper plate in a one-sided submerged-arc welding (SAW).

(c) *Adaptive control based on an optical sensor*

A system is available which controls the width of back beads and the height of weld beads based on the application of a point sensor. This system adopts four CdS sensors which detect the arc light quantity of SAW penetrated into the back of the plate, and keeps the width of the back beads constant by controlling the welding current so that a constant amount of arc light may be obtained. Furthermore, in this system, an attempt is made to control filler wire amounts correlated to the welding current and obtain a constant height of weld bead.

Figure 1.11 showed an example of an image-processing adaptive control system based on the application of an area sensor which:

1. detects groove width, and maintains a constant depth of penetration and height of beads (deposit quantity) under the control of welding current and welding speed;
2. detects groove width and maintains a constant width of weld beads and back beads under the control of the polarity ratio of a.c. pulsed MIG arc welding;
3. detects the width of the molten pool and maintains a constant depth of penetration under the control of welding current.

### 1.3.3   Seam tracking control and adaptive control

A recently developed welding control system can perform simultaneously both seam tracking and adaptive control of welding conditions based on information obtained from either a single sensor or combined sensors.

With the single-sensor system, seam tracking and the amount of deposit are controlled by oscillation-reverse control method, using the arc phenomena, or seam tracking and weld bead shape and back bead shape are controlled by image processing of optical or heat distribution, using an optical area sensor.

With multiple sensors, seam tracking and bead shape are controlled based on either a combination of arc sensor and a linear and optical sensor, or a combination of arc sensor and area (image) sensor.

### 1.3.4   Weld monitoring

Weld monitoring is intended to perform on-line supervision of the welding process, but it may also be used for simultaneous seam tracking or adaptive control of welding conditions. The detectors used for monitoring include: (a) an ITV camera which captures an image; (b) a microphone which monitors arc sound; or (c) a combination of (a) and (b) (equivalent to sight and hearing out of the five human senses).

## REFERENCES

1. Masumoto, I., Araya, T., Iochi, A. and Nomura, H. (1983) Development and application of sensors and sensor systems for arc welding. *Journal of the Japan Welding Society*, **52** (4), 339–47.
2. Murakami, E. and Kugai, K. (1990) Recent arc sensor for arc welding robots. *Journal of Japan Welding Engineering Society*, **38** (1), 120–26.
3. Nomura, H. and Sugitani, Y. (1989) Present new sensors for arc welding. *Journal of Joitech*, April, 59–65.
4. Inoue, K. (1982) Sensors for arc welding robots. Journal of the *Japan Welding Society*, **51**, (9), 735–41.

# CHAPTER 2

# State-of-the-art review of sensors and their application in arc welding processes

## 2.1 CONTACT SENSORS

### 2.1.1 Sensors used in seam tracking

#### (a) Contact probe

Figure 2.1 shows the structure of a typical contact probe. This structure is designed to transmit the displacement of a contact probe in a groove to the other end by way of a fulcrum. It opens and closes microswitch contacts when the displacement deviates from a value which exceeds a dead zone in either direction. The displacement is taken out as a contact signal which drives an actuator (usually an electric motor), moves a welding torch

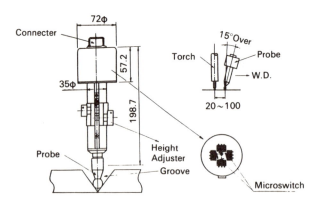

**Fig. 2.1** Contact probe sensor.

*Sensors and Control Systems in Arc Welding.* Edited by Hirokazu Nomura. Published in 1944 by Chapman & Hall, London. ISBN 0 412 47490 5

(mounted with the sensor in one unit) to the central point (dead zone) and corrects its position [1].

Almost all the sensors used in Japan are designed for two-axis control service and are equipped with two pairs of fulcrum and microswitches as illustrated.

### (b) *Memory playback type* (*sensor preceding type*)

When a contact sensor is used, a standard sensor is positioned ahead of the welding torch. A playback type is capable of tracking the weld seam in every shape, even though the sensor precedes the welding torch.

One of the methods of achieving this is called the *memory delay playback system*. An $x-y$ drive block allows the sensor to trace the groove while a separately installed $x-y$ block (block capable of moving vertically and horizontally) adjusts the position of the welding torch. Deflection of the sensor as it traces the joint groove gives a deviation value which is converted into an electric signal and stored temporarily. When the welding torch is moved towards the previous position of the sensor, its position is adjusted by the stored derivative value of the sensor at that point. The welding seam can therefore be tracked satisfactorily (see Figure 2.2).

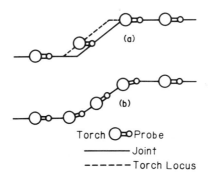

Torch ⬭Probe
———— Joint
—————— Torch Locus

**Fig. 2.2** Sensor system with memory delay playback: (a) standard; (b) memory delay playback.

In addition, a *teaching playback system* is available, which lets a sensor survey the welding seam prior to welding, stores deviation values from the joint and drives the torch by the deviation value when welded in playback mode [2].

### (c) *Sensor whose centre is aligned with the centre of a torch*

When a welding robot is used to weld a complex joint geometry, the previous sensors are subject to some limitations.

The contact sensor illustrated in Figure 2.3 has a ring whose contact section is concentric with the welding torch and provides unlimited welding

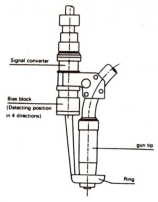

**Fig. 2.3** Welding gun with ring transducer for position recognition.

direction. When carrying out fillet welding or V-groove butt welding, the sensor is used so that it may be pushed into contact with both members of the joint. The detection method is similar to that of the other contact sensors [3].

(d)   Electrode contact sensor

This method has been developed for arc welding robots. It effectively detects deviation between a taught point in the robot path and the actual position of the joint at that point, or poor assembly accuracy. A second routine is taught as part of the robot program. This brings the welding electrode wire projecting from the torch into contact with the component substances. The contact spot is confirmed by applying a voltage of 400 V at 600 Hz to the welding wire and detecting the flow of a small current induced by the contact [4]. This detection method comprises a two-point method which detects one point on each of two right-angled wall surfaces and calculates and positions the wire tip to bisect the right angle (refer to Figure 2.4). A

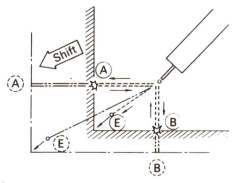

**Fig. 2.4** Wire touch sensor.

four-point method can also be used for other surface intersection angles or for detecting the end of the component.

### 2.1.2 Sensors used in torch angle control

When an attempt is made to weld a corrugated plate or a steel pipe in automatic mode, it is necessary to control a welding torch so that it maintains a right angle to the material surface. To comply with this, there is a demand for a sensor capable of detecting the angle made by a welding line (or surface) and a welding torch.

Figure 2.5 shows a sensor developed and applied for lap welding of a corrugated plate. The sensor is equipped with two styluses. They are laid out slightly separated from each other in the welding direction. When the sensor (integral with the welding torch) approaches an inclined portion, a differential elevation $\Delta d$ is produced between the preceding stylus 2 and the following stylus 1 as illustrated in Figure 2.6. This $\Delta d$ is converted into a voltage by the potentiometer built in the sensor. A torch rotation mechanism makes it possible to maintain the torch at a right angle by slanting the torch so that this output voltage is zero [5].

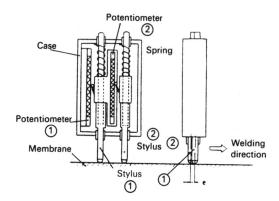

**Fig. 2.5** The contouring sensor.

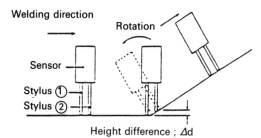

**Fig. 2.6** Detecting an inclined surface.

### 2.1.3   Sensors used in detecting groove width

Figure 2.7 shows an example of a sensor which detects a groove cross-sectional area, using the electric contact sensor described above. The roller of a groove-tracing sensor (1) is inserted into a groove while sensor (2) is used simultaneously to compensate for the deformation of the materials [6]. The groove width and cross-sectional area are calculated from the differential depth of sensors (1) and (2) and the groove angles $\theta_1$ and $\theta_2$.

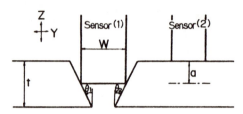

**Fig. 2.7** Groove width detection sensor.

## 2.2   TEMPERATURE SENSORS

Non-contact temperature sensors can be used to detect infrared light irradiated from an object or to measure irradiated power. Applications to control welding phenomena are limited at present, but considerable research is in progress to extend the use of such sensors.

If the object is regarded as a black body, then the temperature could be determined by measuring the radiated power of a specified wavelength of radiation spectrum $L(\lambda, T)$, the intensity ratio at two different wavelengths, or the wavelength that indicates the maximum intensity. However, in practice the measurement object is not a black body, and therefore the radiation energy becomes $L'(\lambda, T) = \varepsilon(\lambda, T) L(\lambda, T)$. The emissivity $\varepsilon(\lambda, T)$ varies with a substance or its surface state, temperature and wavelength. There are two types of radiation pyrometer used in practical applications: monochromatic temperature pyrometers designed to measure radiation energy in a single wavelength, and colour temperature pyrometers designed to obtain temperatures from the radiation energy ratio in two different wavelength bands. The structure of the monochromatic type is illustrated in Figure 2.8 [7].

Figure 2.9 shows the sensitivity of various infrared semiconductor sensors. Only Si sensors are normally used at room temperature [8]. Figure 2.10 shows an application of Si as a monitor to detect the temperature on a welding line. This sensor is used for TIG pulse (peak current: 300 A, base current: 5 A) welding. A signal transmitted from a Si sensor undergoes a marked change when it is subject to arc noise, but a measured value during base current is virtually identical to that detected by a thermocouple except for that in the front of a molten pool [9].

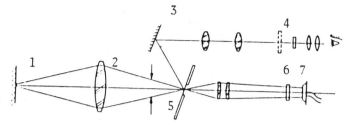

**Fig. 2.8** Optical alignment of monochromatic temperature pyrometer: 1, object; 2, object lens; 3, mirror; 4, grey filter; 5, slit and aperture mirror; 6, interference filter; 7, sensor (Si).

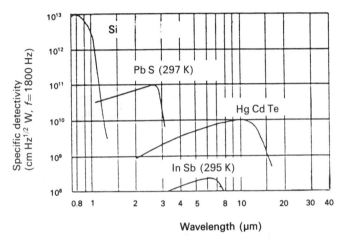

**Fig. 2.9** Specific detectivity of some infrared light sensors.

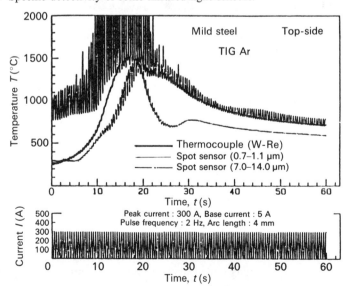

**Fig. 2.10** Temperature measurement using a photosensor.

A measuring instrument has been developed which comprises a semi-conductor infrared area sensor capable of capturing two-dimensional temperature distributions on a CRT screen. Depending upon the application requirements, this sensing system can serve as a welding process monitor or as the basis for process control. It can also be used to obtain the temperature distribution of a molten pool, the heat affected zone (HAZ) or the temperature distribution in a plate on the back side [10, 11]. As illustrated in Figure 2.11, the positional deviation of arc, faulty joint geometry, change in bead width, and impurities on the surfaces can be recognized by the analysis of the thermal image. In addition, it is possible to obtain the cooling rate of the weld metal (Figure 2.12) or the shape of the molten pool from the temperature distribution and it is also possible to control the welding process. Feed position control of filler wire is carried out based on this system for narrow groove hot wire TIG welding [12].

**Fig. 2.11** Thermograph of TIG weld pool.

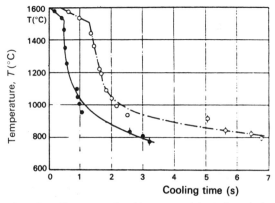

**Fig. 2.12** Examples of cooling rate of weld measured by infrared thermography.

## 2.3 ARC PHENOMENA SENSORS (THROUGH-THE-ARC SENSORS)

### 2.3.1 Arc sensor (seam-tracking sensor)

The arc sensor eliminates the use of separate detector equipment around a welding torch. It can detect groove position directly or provide information on the surface position of a molten pool under an arc almost in real time, and can therefore be used as a seam-tracking sensor for arc-welding robots applied to structures having many internal corners. This is why the arc sensor is currently the most popular of the various welding sensors.

#### (a)  *Principle of the arc sensor*

Figure 2.13 shows the static arc characteristics of an MAG arc using a mild steel wire and a constant voltage characteristic power source, and shows the change in the distance between the wire tip and the base metal at a constant wire-feeding speed. The welding current will be increased when the distance between the tip and the base metal is decreased; the arc voltage will be decreased.

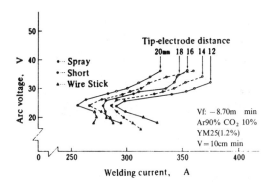

**Fig. 2.13** MAG arc characteristics.

The change $dI/dEx$ in welding current $dI$ against the change in distance between tip and base metal, $dEx$, can be approximated as

$$\frac{dI}{dEx} \approx -\frac{BI^2}{A+2BLI} \tag{2.1}$$

Equation [1] is derived from Lesenvich's formula relating to wire melting rate where $A$ and $B$ are the constants for wire and shielding gas while $L$ is wire extension length. Equation [1] clearly indicates that the current change rate (voltage change rate) is increased as the welding current and wire feeding speed are increased while the wire extension length and wire diameter are decreased.

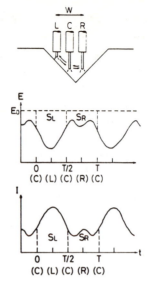

**Fig. 2.14** Current and voltage waveforms in an oscillated arc.

Figure 2.14, which shows the waveforms of the welding current and voltage when the arc is oscillated across the groove, is based on the characteristics illustrated in Figure 2.13.

The sensitivity of the arc sensor is greatly affected by the oscillation frequency as well as the parameters given in Equation [1]: the higher the oscillation frequency, the greater the available sensitivity. Figure 2.15 shows

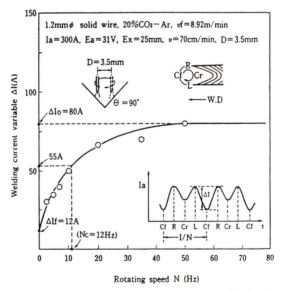

**Fig. 2.15** Relationship between arc rotation speed and amplitude of welding current waveform.

the relationship between the rotation speed and the amplitude $\Delta I$ of the welding current waveform obtained when an electrode nozzle with an eccentric wire guide is rotated [13]. If the rotation speed is high, self-regulation of the arc does not operate and the wire extension length does not change. In this case, the current change rate is represented by

$$\frac{\mathrm{d}I}{\mathrm{d}Ex} = \frac{X}{\alpha + \beta} \tag{2.2}$$

where $X$ is the potential gradient of arc column, $\alpha$ is the inclination of the arc characteristics and $\beta$ is the drooping inclination of the power output characteristics [13].

### (b) Application to seam tracking

Figure 2.16 shows a method in which the peak values (at each peak half-cycle of the wire oscillation) of welding current waveforms are compared. This method is the most popular and hence has the largest number of application examples [14, 15]. Figure 2.14 showed the method where the waveforms in the left and right half-cycle of the oscillation are compared; it is applied to narrow gap welding, for example [16]. This method is also applicable to short-circuit transferred arc welding including several 10 Hz ripples in the waveforms [17]. Recently, many other methods with arc oscillation have been developed. They include a processing method [18] that takes out components of the frequency equivalent to an oscillating frequency and the components of the double frequency, and a fuzzy control-based method which calculates the distance between the tip and the material in-process and estimates the joint shape [19].

Since the above methods are based on oscillating the welding wire mechanically in reciprocating mode, the oscillation frequency ranges from 4 to 5 Hz are mechanically, and hence the application to high-speed welding or lap welding for thin sheets is limited. Figure 2.17 shows the mechanism for the high-speed rotating arc welding method, which makes it possible to realize rotary oscillating motion of around 100 Hz. As shown in

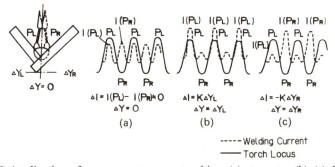

**Fig. 2.16** Application of arc sensor to seam tracking: (a) at centre; (b), (c) deviated.

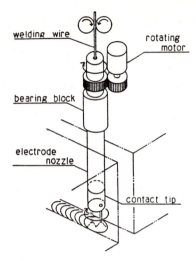

**Fig. 2.17** High-speed rotating arc.

Figure 2.15, the detection sensitivity can be improved so that the application limitations can be reduced dramatically [20].

In the above cases, the torch height control method is adopted, which compares the mean value of the welding currents with the reference value.

### (c)   *Application to seam tracking and deposition*

When arc length control is adopted with sufficient response speed and oscillation of the torch in the groove, the torch will trace a locus parallel to the groove surface as shown in Figure 2.18. In this case the displacement of the torch in the $y$ direction is detected, and when this detected value reaches the reference value ($e_y = e_0$), the oscillating direction is reversed. This method can not only trace the groove but can also control the oscillating motion so that an optimum oscillation width may be obtained to compensate for irregularity of the width or depth of the groove.

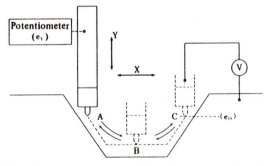

**Fig. 2.18** Principle of arc sensor seam tracking (constant arc length control).

The oscillation locus of the torch also provides information about the height of the molten pool under the arc. Figure 2.19 shows an example of deposition control. Figure 2.19(a) indicates the initial state of a groove; Figure 2.19(b) shows a change in the groove width. Width and area of the oscillation locus are changed from $x_0$ to $x$ and from $S_0$ to $S$ respectively. To maintain a constant height of bead, the cross-sectional area of deposit should be increased by $\Delta S$ as shown in Figure 2.19(c), where

$$\Delta S = S - \frac{x}{x_0} \cdot S_0 \tag{2.3}$$

The welding speed $v$ and wire-feeding speed $v_f$ are controlled based on

$$v_f = C(A + \Delta S) \tag{2.4}$$

$$A = \frac{v_{f0}}{Cv_0} \tag{2.5}$$

This method is applied in TIG welding [21].

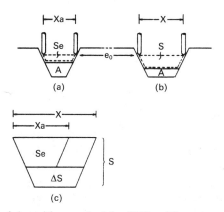

**Fig. 2.19** Principle of deposition control for TIG welding.

Figure 2.20 shows another example of deposition control based on a similar seam-tracking method. It is used for $CO_2$ and MAG welding, where the difference between the oscillation width $W$ and the initial value $W_0$ in every oscillation cycle is almost the same as the change $G - G_0$ in the groove width. The welding speed is therefore controlled so that the cross-sectional area of the deposition may change by $h_0 (W - W_0)$ where the height of bead on one layer is assumed to be $h_0$. This welding method is applied to multilayer welding for thick plates, as in steel fabrication since it is applicable to each layer of multipass welding [22]. The relevant equation is:

$$v = \frac{v_f}{A_0 + \Delta A_0} \tag{2.6}$$

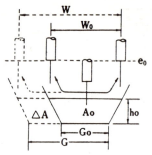

**Fig. 2.20** Principle of deposition control for $CO_2$ and MAG welding.

where $v_f$ = wire feeding rate, $v_0$ = initial welding speed, $A_0 = v_f/v_0$ and $\Delta A = (W - W_0) h_0$.

### 2.3.2  Sensor used in back bead control in one-sided welding

The multi-electrode one-sided automatic welding method is a popular welding technology in the shipbuilding industry. In this method, a key hole is formed under the leading electrode arc, and if the back beads are formed correctly, the arc is visible at the back side of the plate.

Figure 2.21 shows a one-sided SAW back bead control method which utilizes the relationship between the quantity of arc light passing through to the back side and the back bead width. Four CdS elements for arc light intensity detection are embedded on a water-cooled copper backing strip which can slide on the back of the weld. The welding current is controlled so that the arc light intensity maintains a preset reference value. It is necessary to drive the backing copper strip synchronously with the welding arc. Control of the running motor is based on the difference between the front two CdS outputs ($F_R$ and $F_L$) and the rear two ($B_R$ and $B_L$) [23].

In FCB one-sided SAW, where fluxes are distributed to a fixed copper plate so that it may serve as a backing strip (Figure 2.22), a potential difference of several volts is detected between the copper backing strip and

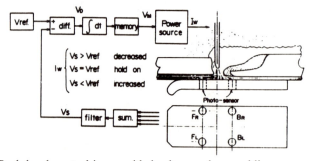

**Fig. 2.21** Back bead control in one-sided submerged arc welding.

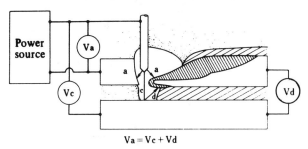

$$Va = Vc + Vd$$

**Fig. 2.22** Arc voltage distribution in one-sided submerged arc welding.

the base metal. In this case, there is a good relationship between the detected voltage and the back bead width. The welding current is controlled to keep the detected voltage constant [24].

When stabilized back beads are formed by controlling the arc light intensity in the control system described above, there is a good relationship between the controlled welding current value and the gravity head of the molten pool. This relationship has been developed into a control system in which, at the same time as back bead width control, bead height is constantly controlled by controlling the amount of filler wire, so that the welding current is kept constant in relation to the bead height determined by the plate thickness [23].

One-sided welding with the $CO_2$ (MAG) process is usually carried out with a root gap of several mm. Several control systems have been developed based on this welding method [25]–[28]. Figure 2.23 shows the method which forms the optimum back beads by controlling the welding speed so as to control constantly the number of short circuits in die-transfer welding. This utilizes the correlation between the distance from the front edge of a molten pool to the arc generation position, back bead formation state, and the number of short circuits [25].

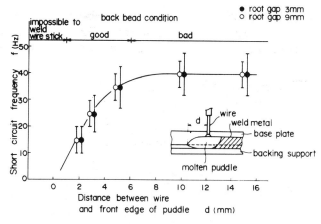

**Fig. 2.23** Correlation between short-ciruiting frequency and wire position.

## 2.4 OPTICAL SENSORS

### 2.4.1 Characteristics of optical sensors

#### (a) Optical point or linear sensor

In one application, an optical sensor is used to detect the position of the end face of a spirally welded steel pipe. A projector lamp and a light receiver are arranged to face each other. A steel plate is clamped at the end to close the steel pipe, and the optical element (photodiode) is used to detect the amount of light which reaches the receiver. The optical sensor is then repositioned so that the amount of light received is fixed. Figure 1.11(a) shows the application of a point sensor in which a photodiode is oscillated or rotated so that it crosses the welding line. The amount of light received determines the position of the sensor in two dimensions.

As illustrated in Figure 1.11(b), the principle of a linear sensor is to obtain the position of a reflected point by projecting the light from a point light source to a groove, receiving the reflected light by linear sensors (CCD, MOS and PDS), and detecting $\Delta h$. Figure 2.24 shows a practical example. This is a sensing system in which the light from a semiconductor laser diode is scanned across the surface of the joint. The reflected light is received by a position-sensitive detector (PSD) so that the height of the sensor from the surface may be detected, enabling calculation of joint angles, root face thickness, groove gap, a groove cross-sectional area, and the central position of a groove. When used to scan a completed weld, the height of the weld bead and weld profile may be determined [29]. Figure 2.25 shows an optical sensor mounted on a welding torch, which is designed to detect light from a semiconductor laser with a linear array of photodiodes to trace the groove and identify the joint geomtry [30]. The accuracy and resolution of this

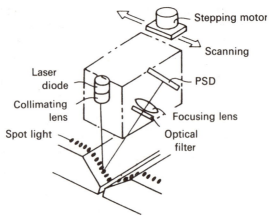

**Fig. 2.24** Laser sensor using PDS.

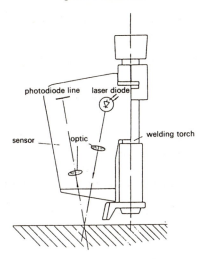

**Fig. 2.25** Optical sensor mounted on welding torch using a line of photodiodes.

sensor depends on the design of the optical system and diode array. The accuracy is $\pm 0.1$ mm, which is sufficiently high. The detection time for torch position is as short as 25–30 ms.

### (b)   Area sensor using a camera

This is an area sensor equipped with a camera having a picture element comprising a large number of optical elements (for example, $256 \times 256$ CCD or MOS type) arranged in the $x$ and $y$ directions to detect an object visually. These enable large amounts of information to be obtained by processing of the received image. However, the disadvantages of such sensors is that they require a slightly longer time for image processing. A CCD camera or industrial TV (visual control camera) can be used as an area sensor. Compared with the visual control camera, the CCD camera (or an MOS-type camera) is smaller in size, lower in weight, and highly resistant to electromagnetic noise. This is why this method is so popular. Using a 16-bit computer, which has replaced the 8-bit computer, dramatically reduces the image-processing time.

The laser visual sensor, which uses a semiconductor laser or He–Ne laser as its light source, operates on the light-cutting method illustrated in Fig. 1.11(c): it is very popular as a means to detect joint position or configuration [31]. This system is designed to operate by projecting a line of laser rays or a stripe of light on the weld, receiving the reflected light with a CCD camera, and converting the two-dimensional information into three-dimensional information by image processing. This method is very effective for shape recognition and hence it is frequently used to detect not only groove configuration [32], but also the leg length of fillet welded joints, and bead width [31]. This system is built into the arc welding torch [32].

One method for obtaining a stripe of light is to use a cylindrical lens. An alternative method oscillates a mirror in a light channel. An area sensor system is also available in which a fibre scope is adopted. The light from the He–Ne laser light source illuminates the object to be welded at a specific angle via an optical fibre. The reflected light enters the area sensor or a light receiver where optical elements are laid out in two dimensions, and then detected. This sensor is capable of detecting both longitudinal and transverse directions of the arc [34]. The method can trace welding lines with high accuracy. The sensor has the advantages of being low in weight, small in size and compact in installation space. However, the number of pixels is limited, which degrades the resolution, darkens, and degenerates the image easily.

### 2.4.2   Application of optical sensors and sensing systems

Table 2.1 shows the applications of optical sensors, which involve control of welding torch position or orientation in space and welding process control. These two functions are used simultaneously in a fully automated welding system.

**Table 2.1** Applications of optical sensors

---

Position control and shape recognition
　　Control of torch position (seam tracking)
　　Control of wire tip deviation
　　Control of TIG electrode position
　　Recognition of molten pool and arc configuration

Welding process control
　　Control of penetration depth by detecting the width of molten pool
　　Control of bead configuration by detecting quantity of weld metal
　　Control of filler wire feeding speed in TIG process
　　Simultaneous control of back bead width in MIG seam tracking in real time

Simultaneous welding process control and seam tracking
　　Adaptive control of arc welding robot
　　Full automation of remote-controlled TIG process
　　Full automation of submerged arc welding

---

During control based on these optical sensors, it is necessary to resolve disturbances such as welding arc or electromagnetic noises. Recent research and development has made progress in solving these problems so that such disturbances can be controlled on line. More specifically, many new design concepts have been introduced for the short-circuiting arc welding and pulsed arc welding processes. They include:

1.  capture of image data during short-circuiting or the base current period;
2.  determination of the wavelength of luminescent elements based on an optical filter;

3. balancing the amount of light by rotating a slit disk in front of a TV camera;
4. reduction of the effect of fumes and spatters by allowing part of the shielding gas to flow in front of the camera lens.

Examples of processing optical information detected by optical sensors are described below.

### (a) Observance of iron spectrum in TIG arc (monitoring of gas-shield disturbance)

Figure 2.26 shows a sensor system which uses a spectroscopic approach to monitor the flow of air into the shielding gas [35]. Spectrograph analysis shows that an argon line spectrum is radiated in standard TIG welding, but a large number of iron spectra are generated if air is injected into the arc. This makes it possible to monitor the state of the shield by observing the spectral lines of the iron. Based on the experimental results, it is necessary to adopt an iron spectrum of around 523 nm in consideration for the spectral intensity. The arc light is sampled by the application of a quartz rod, introduced to a spectral analysis system through an optical fibre, and filtered by an interruption filter (half value width 2 nm) where the 523 nm line is selected. The arc light is converted into an electrical signal by a semiconductor sensor. In a signal-processing system, a band filter (50–100 Hz) separates the input signal into the a.c. component ($S$ value) equivalent to the intensity of pulse-like iron spectral lines while a low-region filter (0–10 Hz) separates the signal into the d.c. component ($N$ value) equivalent to the change in the quantity of light. Then, an attempt is made to obtain $S/N$ with a divider to select the intensity of the iron spectral lines irrespective of the fluctuations in the quantity of light induced by weaving, and current variations. When this value exceeds an allowable value, a comparator operates so that an alarm sounds.

### (b) CCD camera-based observation of molten pool phenomena and groove status

When the arc light is strong, the molten pool is relatively dark. Therefore, if it is necessary to monitor the molten pool with a CCD camera, an attempt

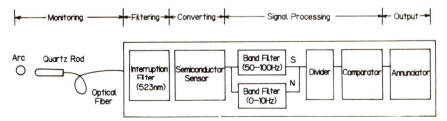

**Fig. 2.26** Monitoring device.

must be made to lower the welding current periodically for a short time to reduce the intensity of the arc light, and synchronize the camera shutter with this period. Alternatively, the weld pool may be monitored during the short-circuiting period if short-circuiting metal-transfer conditions are being used.

### (i) Pulse TIG MIG MAG welding

Figure 2.27 shows the block diagram of this system. Figure 2.28 shows the timing chart for the observation of a molten pool. Camera 1 is attached to detect the welding line and groove while camera 2 is attached to observe the molten pool. The camera synchronizes with a vertical synchronizing signal $V_D$ and capture one static image per 1/60 s. However, since processing of the weld pool image takes time, an image is only captured every fourth cycle (4/60 s). To capture a clear image of the surface of a molten pool and joint groove clearly, it is necessary to reduce the current to 15 A, synchronizing with the shutter of the camera (1 ms). The molten pool image and groove image will then be displayed on the screen of a television monitor.

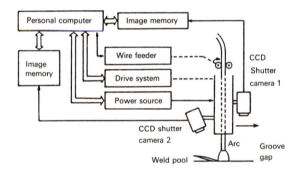

**Fig. 2.27** Groove gap detection and weld pool control system.

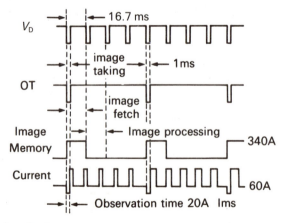

**Fig. 2.28** Timing chart.

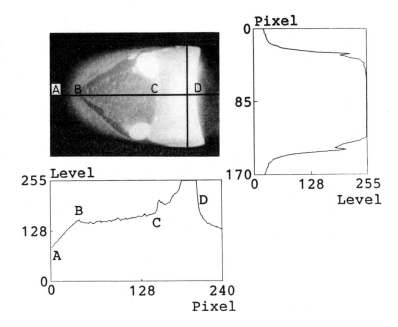

**Fig. 2.29** Image of weld pool.

Figure 2.29 shows the image of the molton pool captured with camera 2 and the average brightness distribution of ten vertical picture elements at each position on the straight line A–D. The points A, B, C and D on the straight line A–D correspond to those of the brightness distribution while the point B corresponds to the solid liquid boundary point. By obtaining the brightness level in the molten pool boundary at point B, it is possible to obtain the width of the molten pool at each part, defining the part which exceeds this brightness level as a molten pool [36].

**(ii) Image processing of shot-circuiting arc welding**
As shown in Figure 2.30 an attempt is made to detect the period when the voltage $V$ of the welding torch at the current supply point is turned to a low voltage (15 V and lower) as a short circuit period. External trigger pulses are then generated and transmitted to the CCD camera by a timing pulse generator. The shutter of the camera is opened by the last transition of the trigger pulses (shutter time 0.5 ms) so that at each short-circuit period the molten pool may be captured by camera 2 while simultaneously the groove configuration may be captured by camera 1. Figure 2.31 shows a typical molten pool image of short-circuiting welding captured by the CCD camera [37].

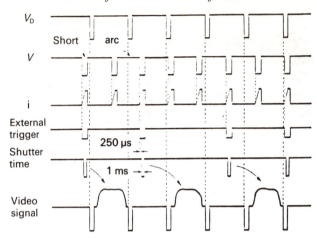

**Fig. 2.30** Timing chart in short-circuiting arc welding.

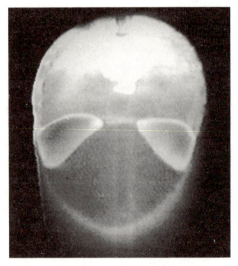

**Fig. 2.31** Image of weld pool in short-circuiting arc welding.

*(c)  Simultaneous control of seam tracking and welding conditions (sensor system for full automation)*

This system is designed to detect and control groove configuration and the positions of wire and torch. More specifically, in addition to seam tracking, control and observation of the molten pool and the welding process are image-processed and analysed, and processed by the introduction of AI (artificial intelligence) and an expert system, thereby creating a more advanced fully automated welding system. An application to narrow groove GMA has been demonstrated [38]. Many attempts are being made to develop an automatic welding system which is capable of monitoring (a) deflection of wire, (b) arc state, (c) penetration situation in the base metal, and (d) detection of problems, and then processing the information with AI, so that the system can compensate for such factors as work deformation, dimensional error and welding condition.

## 2.5   ELECTROMAGNETIC SENSORS

The electromagnetic sensor is a non-contact type which uses an electromagnetic field as a medium. It is not used for welding as frequently as might be expected. However, it is very simple in construction and hence widely used for the detection of mechanical variables (such as position or displacement), the measurement of biomagnetic field or remote sensing of underground resources for surveying. Electromagnetic sensors used to detect position or displacement are classified into capacitance types and eddy current types. The former are greatly affected by the flatness or parallelism of the surface to be detected, and are therefore occasionally used in cutting, but seldom adopted for welding.

When a magnetic flux which interlinks a conductor is subject to change, an eddy current is generated in the conductor. As illustrated in Figure 2.32, if the distance between a flux generation coil and a conductor is changed, the magnitude of the eddy current is changed, thereby changing the impedance of the coil. This is why this function is used as a non-contact displacement detector.

Since the relationship between the impedance of the coil and the distance to be detected is affected by the electrical conductivity and magnetic permeability of the object and the geometrical configuration, it is very difficult to define the distance relationship theoretically. Furthermore, the greater the distance, the smaller the sensitivity for detection of the object. Therefore, it is very important to place this type of sensor near the object. To comply with this, the sensor must meet the reqirements of simple construction, small size and low weight. Furthermore, it must be capable of achieving satisfactory detection sensitivity even if it is placed in

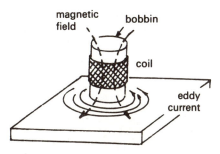

**Fig. 2.32** Electromagnetic sensor.

an environment which is easily exposed to arc heat or spatter. Various design concepts have been introduced to compensate for temperature effects.

Figure 2.32 shows an eddy current sensor which comprises a single coil designed for both excitation and detection, using the changes in the impedance for the high-frequency current of this single coil. Figure 2.33 shows another type, which comprises a magnetic field generation coil and separate detection coils. The detection coils $C_1$ and $C_2$ have the same number of turns and differential connection so as to inhibit the effect induced by the other magnetic field as much as possible [39].

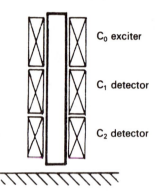

**Fig. 2.33** Example of electromagnetic sensor.

In the single-coil type, a weak d.c. current is superimposed on a high-frequency current, and the drop in the d.c. resistance is used to detect the temperature of the coil and compensate for the thermal effect [40]. As a result, linearity between the distance from the surface and the output voltage can be obtained for distances up to 10 mm. Figure 2.34 shows the basic characteristics of the single-coil sensor.

The sensor head is around 10 mm in diameter and 10–40 mm in length. The excitation frequency is at least 100 kHz to minimize the effect of electric conductivity and magnetic permeability. The detection sensitivity reaches 0.1 μm, but setting sensitivity ranges from 0.5 to 2 V/mm. Positional control is carried out with a sensitivity of 0.1 mm for practical application. In addition, the scope of the measurement distance is 10 mm maximum.

Two sensors have been used to detect the position of an intersection in fillet welding, thereby performing seam tracking, and the system is also applied to butt welding and narrow gap welding. Figure 2.35 shows a block diagram of the basic system.

## 2.6  SOUND SENSORS

A microphone is used as a sensor to detect audible sound. Many attempts are being made to study its feasibility as a sensor to monitor welding

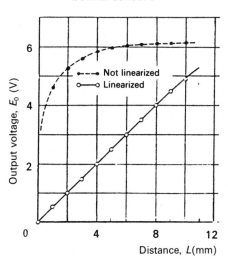

**Fig. 2.34** Basic characteristics of single-coil sensor.

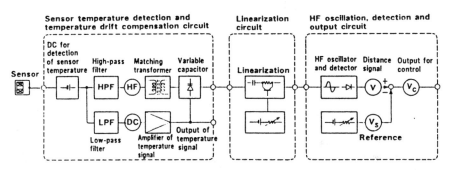

**Fig. 2.35** Comprehensive function block diagram.

phenomena [41], [42]. A microphone is installed at a position 400 mm distant from the arc point (120 mm for TIG welding) so that adequate sound pressure and S/N ratio may be obtained.

The detection factors involve acoustic wave type, spectrum, correlation function (self-correlation, mutual correlation) and periodicity, but currently an attempt is being made to study the correlation between acoustic pressure level in the whole band and welding phenomena. Moreover, attempts are being made to control welding phenomena based on the propagation and reflection of the supersonic wave, but they are still in the R&D stage.

In addition, AE (acoustic emission) is widely used in various kinds of tests for welded structures and in failure predictions. A large number of supersonic sensors have been adopted for practical services. However, this book does not deal with them since they are used to control arc welding phenomena.

# REFERENCES

1. Masumoto, I., Araya, T., Iochi, A. and Nomura, H. (1983) Development and application of sensors and sensor systems for arc welding in Japan. *Journal of the Japan Welding Society*, **52** (4), 339–47 (in Japanese).
2. Nomura, H. (1985) Sensors and sensor systems for arc welding. *Journal of Welding Technique*, **33** (6), 39–43; **33** (12), 74–80 (in Japanese).
3. Heitmayer, U. and Weman, K. (1981) Adaptive sensing: key to the adult robot. *Welding and Metal Fabrication*, December, 592–6.
4. Towata, T. (1979) Outline and its application of Shin Meiwa sensorless sensing system. *Robot* no. 25, 16–20 (in Japanese).
5. Nomura, H., Fujioka, T., Wakamatsu, M. and Saito, K. (1982) Automatic welding of the corrugated membrane of an LNG tank. *Metal Construction*, July, 391–5.
6. Akashi, H., Shibata, Y. and Tamaoki, M. (1984) Development and application of SAW system with microcomputer, IIW document XII-828–84.
7. Hishikari, I. (1984) *Keiso*, **27**, 12 (in Japanese).
8. Morimura, M. and Yamazaki, H. (1982) *Asakura-shoten* (Sensor Technology) (in Japanese).
9. Moriyasu, M., Hiramoto, M., Ohmine, M., Shibata, Y. and Ohji, T. (1987) *Proceedings of the Annual Meeting of the Japan Welding Society*, April, No. 411 (in Japanese).
10. Chin, B. A., Madsen, N. H. and Doodling, J. S. (1983) *Welding Journal*, **62**, 227s–234s.
11. Franz, U., Stuhec, H. *et al.* (1983) Rechnergestutre Auswertung Thermovisueller Untersuchungen. *Schweisstechnik (Berlin)*, **33**(10), 436–9.
12. Bytyci, B., Meyendorf, N. *et al.* Thermovisionsmessungen bei der Dunnblech-schweissung. (1985) *ZIS Mitteilungen*, **27**(6), 683–8.
13. Sugitani, Y., Tamaoki, N., Kobayashi, Y. and Murayama, M. (1989) Seam tracking and deposition control by arc sensor. Preprint of National Meeting of Japan Welding Society, No. 44, 39–41 (in Japanese).
14. Yokoshima, N. (1982) Sensing technique of Komatsu's arc welding robot. *Journal of Labour Saving and Automation*, February, 36–41 (in Japanese).
15. Nagatomi (1982) Sensing technique of arc welding robot Motoman-L10. *Journal of Labour Saving and Automation*, February, 42–6 (in Japanese).
16. Nomura, H. and Sugitani, Y. (1980) Seam tracking method utilizing an arc as a sensor (Report 1). Preprint of National Meeting of the Japan Welding Society, No. 26, 110–11 (in Japanese).
17. Nomura, H., Sugitani, Y., Suzuki, Y., Tamaoki, N., Kobayashi, Y. and Murayama, M. (1986) Automatic control of arc welding by arc sensor system. Nippon Kokan Technical Report, Overseas No. 47, pp. 23–32.
18. Nakajima, M. and Araya, T. (1986) Arc welding sensor for robot. *Proceedings of the 106th Technical Commission on Welding Processes*, SW-1697-86 (in Japanese).
19. Fujimura, H., Ide, E. and Murakami, S. (1988) Application of fuzzy theory for welding seam tracking. *Proceedings of the 119th Technical Commission on Welding Processes*, SW-1881-88 (in Japanese).
20. Sugitani, Y., Murayama, Y. and Yamashita, K. (1990) Development of articulated arc welding robot with high speed rotating arc process, in *Proceedings of the Fifth International Symposium of the Japan Welding Society*, pp. 525–30.
21. Nomura, H., Sugitani, Y. and Tamaoki, N. (1987) Automatic real time bead height control with arc sensor in TIG welding. *Transactions of the Japan Welding Society*, **18**(2), 35–42.

22. Nomura, H., Sugitani, Y. and Tamaoki, N. (1987) Automatic real time bead height control with arc sensor (report 2). *Transactions of the Japan Welding Society*, **18**(2), 43–50.
23. Nomura, H., Satoh, Y. and Satoh, Y. (1982) An automated control system of weld bead formation in one-sided submerged arc welding. IIW Document XII-C-032-82.
24. Nomura, H. (1987) Back bead width control in one side SAW using flux copper backing. *Transactions of the Japan Welding Society*, **18**(2), 18–25.
25. Kitazawa, Y., Nagai, Y. and Watanabe, T. (1986) Through the arc sensing system for GMA one side welding. IIW Document XII-931-86.
26. Ono, H. and Morita, T. (1986) Study on one-side narrow gap MAG welding (Report III) – Development of control system for uranami welding. *Quarterly Journal of the Japan Welding Society*, **1**(4), 17–22 (in Japanese).
27. Sugitani, Y. and Nishi, Y. (1990) Development of portable rail welding robot. Preprints of National Meeting of the Japan Welding Society, No. 47, 118–19 (in Japanese).
28. Nomura, H., Sugitani, Y. and Tamaoki, N. (1987) Development of portable multipass arc welding robot. Preprints of National Meeting of the Japan Welding Society, No. 41 (in Japanese).
29. Shimatani, Y. *et al.* (1985) Automatic welding machine with laser sensor, in *Proceedings of the 103rd Technical Commission on Welding Processes*, SW-1623-85 (in Japanese).
30. Drews, P. and Starke, G. (1985) Development approaches for advanced adaptive control in automated arc welding, in *Proceedings IIW International Conference*, Pergamon Press, Oxford, pp. 115–24.
31. Agapakis, J.E., Wittels, N. and Masubuchi, K. (1985) Automated visual weld inspection for Robotec welding fabrication, in *Proceedings IIW International Conference*, Pergamon Press, Oxford, pp. 151–60.
32. Kuhne, A.H. (1985) Ein Beitrag zur Steuerung und Regelung des Automatisierten Schutgasschweissprozesses und zur Anpassung der Schweissparameter an die Jeweilige Fugengeometrie. Doctoral thesis, Aachen Technical University.
33. Davey, P.G., Barratt, J.W. and Morris, J.L. (1987) New horizons for laser stripe sensors. *Metal Construction*, **19**(12), 688–91.
34. Pan, J.L., Ne, F.D., Chen, W.C., Cao, Q.H. and Wu, M.S. (1983) Development of a two-directional seam tracking system with laser sensor. *Welding Journal*, **62**(2), 28–31.
35. Sugiyama, *et al.* (1984) Development of weld monitoring device in TIG welding, in *Proceedings of Welding Arc Physics Committee of the Japan Welding Society*, No. 84–54b (in Japanese).
36. Ohshima, K. (1989) Observation and control of weld pool phenomena in arc welding. *Journal of the Japan Welding Society*, **58**(4), 44–9 (in Japanese).
37. Kubota, T., Iida, H., Ohshima, K. *et al.* (1989) Image processing and control of weld pool in short-arc welding (2), in *Proceedings of Welding Arc Physics Committee of the Japan Welding Society*, No. 89–726 (in Japanese).
38. Watanabe, H., Kondo, Y. and Inoue, K. (1988) Automatic control of narrow gap GMA welding, in *Proceedings of Technical Commission on Welding Processes, Japan Welding Society*, SW-1882-88 (in Japanese).
39. Takano, T. *et al.* (1975) *The Hitachi Hyoron*, **57**, 825 (in Japanese).
40. Wakamatsu, K. *et al.* (1984) *Mitsubishi Juko Giho*, **21**, 692 (in Japanese).
41. Ohmae, T. *et al.* (1984) *Proceedings of the Annual Meeting of the Japan Welding Society*, 35, 218 (in Japanese).
42. Futamata, M. (1980) Thesis (in Japanese).

# CHAPTER 3

# Control systems

## 3.1 SEAM TRACKING AND DEPOSITION CONTROL

Figure 3.1 shows the block diagram of a seam-tracking system for pulse MAG welding. This system is designed to control the height of the torch by oscillating the torch in a groove. The welding power source for pulse MAG welding has a constant voltage characteristic. Therefore, when the height of the torch from the base metal is reduced, the arc length is also reduced while the welding current is increased. Conversely, when the height of the torch from the base metal is increased, the arc length is also increased while the welding current is decreased. Based on these features, the height of the torch and the welding current can be controlled. The welding current is measured by the application of a sensor. If the deviation $i_R - i$ is positive, i.e. the welding current is smaller than a reference value, it is necessary to control the $y$-axis motor so that the torch is lowered. In this manner, the height of the torch is controlled with the feedback of current. A potentiometer is used

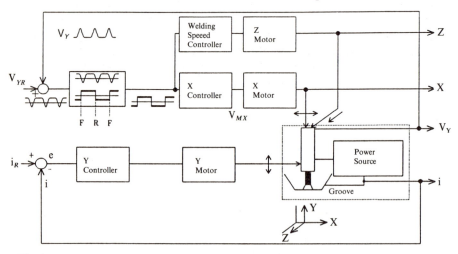

**Fig. 3.1** Control system for wire melting and tracking of groove gap.

*Sensors and Control Systems in Arc Welding.* Edited by Hirokazu Nomura.
Published in 1994 by Chapman & Hall, London. ISBN 0 412 47490 5

to measure the current torch height $V_Y$. When the measured value exceeds the predetermined torch height $V_{RY}$, it is assumed that the torch has moved to the side of the groove. Therefore, the x-axis motor is controlled until the opposite side of the groove is reached. The width of groove can be calculated from the pulse cycle of the signal. As the groove widens, the pulse period becomes longer, which means that the quantity of deposit should be increased, and the z-axis motor should be controlled. Figure 3.2 shows the displacement of x and y axis and the waveform of the welding current under seam-tracking control.

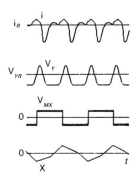

**Fig. 3.2** Displacements and current waveform.

## 3.2   WELDING PROCESS CONTROL

Generally, in robotic welding, the torch position and torch angle on the welding line are taught to a welding robot, as well as the welding parameters, including the welding current waveforms, welding speed, wire feeding speed and welding conditions. They are taught to the robot as preset command data and may be established from the basic experiments made to obtain favourable welding results and quality.

However, when the boundary conditions or system parameters change as the width of the groove fluctuates or the heat conduction varies, the controlled variables, such as molten pool shape, may deviate from the reference value. It is therefore necessary to sense and then control penetration shape or cooling rate so that they conform to the reference values.

As an example of automatic and intelligent performance, the following sections describe the use of weld pool width to control penetration configuration and cooling rate.

### 3.2.1   Sensing of the molten pool and development of the state equation

At first, a basic steady-state experiment was performed in argon gas with 2% of oxygen, using a mild steel wire of 1.2 mm diameter and a 3.2 mm thick plate for the base metal. Figure 3.2 shows the pulse current waveform applied for the experiment. Under the control of the base current, the wire feed rate was varied according to the magnitude of the current so as to obtain an arc length of 5 mm from the surface of the molten pool. Figure 3.3 shows the relationship between pulse current $I$, the width of each part of the molten pool $W_i$, and the penetration depth $D$, where $W_i$ ($i = 1$ to 4) are the widths of the molten pool at positions with distances of 4.5, 6.0, 7.5, and 9.0 mm from the arc centre. The width $W_i$ is generally approximately proportional to the current, but back beads are obtained in a current range from 210 A to 230 A where the width is almost constant with the increase in the current. When the penetration depth approaches the thickness of the plate, the weld pool width changes greatly with change of current, as illustrated in Figure 3.3.

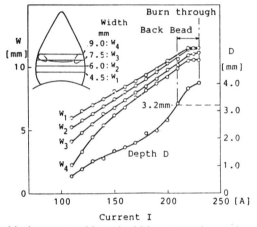

**Fig. 3.3** Relationship between weld pool width, penetration and weld current.

Figure 3.4 shows the transient response of the width of each part $W_i$($i=1$ to 4) when the electric current $I$ is subjected to a step change from 180 A to 220 A. It can be approximated to the response of the first-order lay system. Now as each width $W_i$ and current $I$ is defined as a state variable and an input variable, respectively, the following state equation can be obtained:

$$\frac{dW_i}{dt} = -\lambda_i W_i + b_i I - c_i \qquad (i=1 \text{ to } 4) \qquad (3.1)$$

In accordance with the response of the width $W_i$ to step current, the time constant $T_{ci}(=1/\lambda_i)$ for each value of $i$ is found to be 0.4, 0.5, 0.7 and 0.9, respectively, and the response is slower at the rear part of the molten pool. The parameters $b_i$ and $c_i$ can be calculated from the results of the experiment illustrated in Figure 3.3.

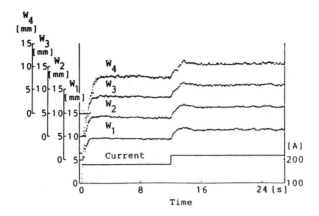

**Fig. 3.4** Transient response of each weld pool width.

### 3.2.2 Back bead control based on molten pool width control

This section discusses the use of molten pool width control to obtain a stable penetration depth (back bead) irrespective of fluctuations in the heat conduction to the base plate fixture. In order to obtain the back bead, the width $W_2$ is required to be uniformly 11.5 mm. Hence the back beads can be adjusted by controlling the width $W_2$.

The state feedback is performed to obtain a desired response. In addition, a welding pool width controller is designed to prevent the generation of offset from a reference value in spite of any change in the system parameters, such as heat conduction. Details will be explained in Chapter 4.

### 3.2.3 Experimental result of back bead control

Two copper plates (100 mm in length × 100 mm width × 10 mm thickness) were brought into contact with the back of the base metal to change the heat conduction. The gap between the copper plates was 10 mm, and the welding experiment was performed with bead on plate. Figure 3.5 shows the response of each width $W_i$ with a constant welding current of 220 A. The change in the width $W_1$ is very small in the section connected with copper plates, but $W_2$ to $W_4$ show a reduction. The maximum reduction is 12% at the width $W_2$.

Figure 3.6 shows the width response at each part while an attempt was made to control the width $W_2$ so that it might be fixed at 11.5 mm. When the manipulated variable current is controlled, the change in the widths $W_1$ and $W_3$ is small regardless of the contact with a support. To clarify further, Figure 3.7 (1)–(4) shows respectively the penetration shape at points (1), (2), (3) and (4) illustrated in Figures 3.5 and 3.6. With constant current, the central part (2) is 2.7 mm deep and no back bead was formed. In contrast

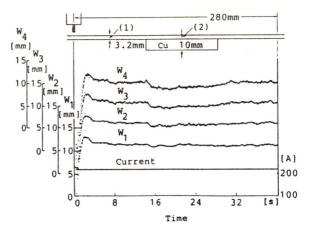

**Fig. 3.5** Time response of weld pool width in constant current.

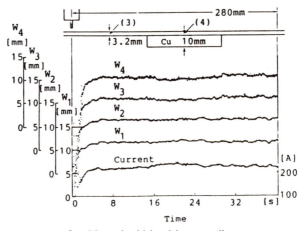

**Fig. 3.6** Time response of weld pool width with controller.

with this, when the width $W_2$ was controlled, the depth was 3.8 mm in the locations with or without a backing plate, and a stabilized back bead formation was obtained over the welded pass.

## 3.3   SEAM TRACKING BASED ON FUZZY CONTROL

This section describes how an arm-type welding robot traces a welding line on a two-dimensional plane. During the operation of the robot, the moment of inertia of each axis would vary. The motion equation is so complicated that it is difficult to apply modern control theory to the system. To solve this problem, therefore, fuzzy control theory, which is based on the experience and know-how of skilful operators, is introduced.

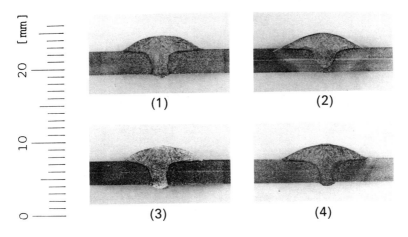

**Fig. 3.7** Penetration shapes corresponding to (1)–(4) in Figures 3.5 and 3.6.

### 3.3.1 Design of fuzzy controller

In Figure 3.8, the deviation $e =$ the distance between the welding line and the torch. The variation $\Delta e$ is selected as the input of the controller, and the advancing angle $\Delta u$ between the torch advancing direction and the line advancing direction is selected for the output. Figure 3.9 shows the fuzzy variables $e$, $\Delta e$ and $\Delta u$.

The control rules will now be considered. If the torch is moved to the left ($u = N$), it will approach the welding line under situation (1) illustrated in Figure 3.10. Then the first rule is:

$$\text{If } e = P \text{ and } \Delta e = P \text{ then } \Delta u = N$$

Figure 3.11 shows the control rules for the situations (2)–(9) illustrated in Figure 3.10. From these control rules, $\Delta u$ is calculated by means of fuzzy inference which will be discussed in Chapter 4. The fuzzy variable

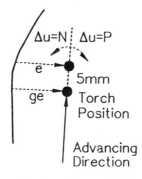

**Fig. 3.8** Relationship between torch position and fuzzy variable.

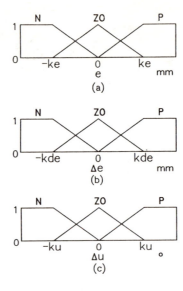

**Fig. 3.9** Fuzzy variables used in tracking of welding line: N, negative; ZO, zero; P, positive.

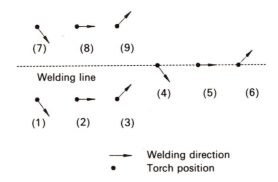

**Fig. 3.10** Various combinations of torch position and advancing direction.

| $\Delta e$ \ $e$ | N | ZO | P |
|---|---|---|---|
| P | ZO | N | N |
| ZO | P | ZO | N |
| N | P | P | ZO |

**Fig. 3.11** Control rules: P, positive; ZO, zero; N, negative.

parameters (*ku, kde, kdu*) shown in Figure 3.9 should be decided with experiments.

### 3.3.2 Seam-tracking experiment

The welding line is detected with a CCD shutter camera, and the torch tip of a welding robot is moved to the start point of a welding line. The forward angle $\Delta u$ was inferred by fuzzy inference according to deviation $e$ and its variable $\Delta e$ obtained from the image memory. The angle of each axis was calculated with the application of inverse kinematics. The experiment was then performed to trace a welding line represented by the fine line in Figure 3.12.

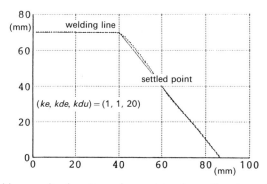

**Fig. 3.12** Tracking result when 5 mm forward torch position is used.

Figure 3.11 shows the result of the tracking experiment where the parameters for the fuzzy variables are $(ke, kde, kdu) = (1, 1, 20)$, and only the information at the position 5 mm ahead of the present torch position was utilized. The settling time was extended. To enhance the tracking performance, an attempt was made to combine the deviation $ge$ at the present

| $\Delta e$ \ $e$ | N | ZO | P |
|---|---|---|---|
| P | ZO | N | N |
| ZO | P | ZO | N |
| N | P | P | ZO |

| $ge$ | N | ZO | P |
|---|---|---|---|
| $\Delta u$ | P | ZO | N |

**Fig. 3.13** Control rules with rule for current torch position added: P, positive; ZO, zero; N, negative.

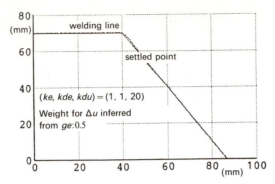

**Fig. 3.14** Tracking result when both current torch position and 5 mm forward position are used.

position and the deviation at the position 5 mm ahead. Figure 3.13 shows the control rules adopted. The weight of 0.005 is attached to $\Delta u$ deduced from the deviation $ge$. As shown in Figure 3.14, settling time is reduced when control is based on the information which covers the present position and the forward position of the torch.

# CHAPTER 4

# Basis for welding process control

## 4.1 MODERN CONTROL THEORY

Discrete signals are handled under digital control with the use of a computer. Figure 4.1 shows the mechanism of a sampler. A switch is closed every $T$ seconds. Let the input signal be the continuous signal illustrated in Figure 4.1(b). If the period when the switch is closed is considerably smaller than the time constant of the control system, the output of the sampler may be considered as a train of impulses (discrete signal $f*(t)$), as shown in Figure 4.1(c).

Since the sampler is switched on every $T$ seconds, the magnitude of the $n$th impulse $f*(nT)$ is identical to that of the input signal $f(nT)$ when $t=nT$. The train of impulses are represented by:

$$f*(t)=f(0)\delta(t)+f(T)\delta(t-T)+f(2T)\delta(t-2T)+\cdots=\sum_{n=0}^{\infty} f(nT)\delta(t-nT) \quad (4.1)$$

$$\delta(t)=\begin{cases}1 & t=0 \\ 0 & t\neq 0\end{cases} \quad (4.2)$$

When $f*(t)$ is Laplace-transformed, the following equation is obtained:

$$L[f*(t)]=f(0)+f(T)e^{-sT}+f(2T)e^{-2sT}\cdots=\sum_{n=0}^{\infty} f(nT)e^{-nTs} \quad (4.3)$$

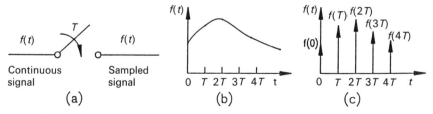

**Fig. 4.1** Sampling process.

*Sensors and Control Systems in Arc Welding*. Edited by Hirokazu Nomura. Published in 1994 by Chapman & Hall, London. ISBN 0 412 47490 5

The simple substitution

$$z = e^{Ts} \tag{4.4}$$

converts the Laplace transform to the $z$ transform. Making this substitution in Equation (4.3) gives

$$Z[f^*(t)] = F(Z) = f(0) + f(T)z^{-1} + f(2T)z^{-2} + \cdots = \sum_{n=0}^{\infty} f(nT)z^{-n} \tag{4.5}$$

where $F(z)$ designates the $z$ transform of $f^*(t)$. Because only values of the signal at the sampling instants are considered, the $z$ transform of $f(t)$ is the same as that of $f^*(t)$.

### 4.1.1 Z transform for unit step function

For this function $f(nt) = 1$ for $n = 0, 1, 2, \ldots$; thus application of Equation (4.5) gives

$$Z[u_s(t)] = 1 + z^{-1} + z^{-2} + \cdots = \frac{1}{1 - z^{-1}} \tag{4.6}$$

This series is convergent for $|z| > 1$. In solving problems by $z$ transforms, the term $z$ acts as a dummy operator. There is no need to specify the values of $z$ over which $F(z)$ is convergent. It suffices to know that such values exist.

### 4.1.2 Z transformation of the exponential $e^{-\alpha t}$

For this function $f(nT) = e^{-\alpha nT}$; thus

$$Z[e^{-\alpha t}] = 1 + e^{-\alpha T}z^{-1} + e^{-2\alpha T}z^{-2} + \cdots = \frac{1}{1 - e^{-\alpha T}z^{-1}} = \frac{z}{z - e^{-\alpha T}} \tag{4.7}$$

The notation $f(n)$ may be used to represent $f^*(t)$. The plot of $f(n)$, shown in Figure 4.2(a), is identical to that of $f^*(t)$. The $z$ transform for $f(n)$ is

$$F(z) = Z[f(n)] = f(0) + f(1)z^{-1} + f(2)z^{-2} + f(3)z^{-3} \cdots \tag{4.8}$$

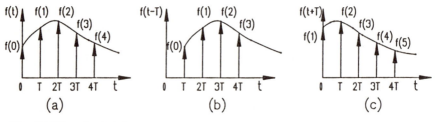

**Fig. 4.2** Discrete signal.

### 4.1.3 Z transform of $f(n-1)$

When $f(n)$ is delayed by one sampling instant, the function $f(n-1)$ shown in Figure 4.2(b) results. The value of $f(n-1)$ when $n=1$ is $f(0)$, the value when $n=2$ is $f(1)$, and so on. The Z transform of $f(n-1)$ is;

$$Z[f(n-1)] = f(0)z^{-1} + f(1)z^{-2} + f(2)z^{-3} + \cdots$$
$$= z^{-1}(f(0) + f(1)z^{-1} + f(2)^z - 2\cdots) = z^{-1}F(z) \qquad (4.9)$$

### 4.1.4 Z transformation of $f(n+1)$

When the function $f(n)$ of Figure 4.2(a) is shifted one sampling period to the left, the function $f(n+1)$ shown in Figure 4.2(c) results. The value of $f(n+1)$ when $n=0$ is $f(1)$, the value when $n=1$ is $f(2)$, and so on. The Z transform of $f(n+1)$ is

$$Z[f(n+1)] = f(1) + f(2)z^{-1} + f(3)z^{-3} + \cdots$$
$$= z[f(0) + f(1)z^{-1} + f(2)z^{-2} + \cdots] - zf(0)$$
$$= zF(z) - zf(0) \qquad (4.10)$$

A schematic diagram of a system controlled by a digital computer is shown in Figure 4.3. At each sampling instant, the digital controller samples the error signal $e(t)$. The controller operates on this sampled value $e^*(t)$ and previous sampled values to obtain an output $u^*(t)$. This value of $u^*(t)$ is then retained by using the zero-order hold until a new value is computed at the next sampling instant. The zero-order hold is illustrated here. The solid

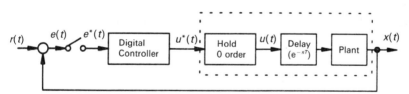

**Fig. 4.3** Digital process control loop.

curve in Figure 4.4 represents the continuous function $f(t)$. The vertical arrows at the sampling instants are the impulses which represent the sampled signal $f^*(t)$. Because the zero-order hold retains the value of $f(t)$ at each sampling instant, $y(t)$ is the series of steps shown in Figure 4.4. The equation for this series of steps (i.e. pulse functions) is

$$y(t) = f(0)[u_s(t) - u_s(t-T)] + f(T)[u_s(t-T) - u_s(t-2T)]$$
$$+ f(2T)[u_s(t-2T) - u_s(t-3T)] + \cdots \qquad (4.11)$$

The Laplace transformation is

$$Y(s) = f(0)\frac{1 - e^{-sT}}{s} + f(T)\frac{e^{-sT} - e^{-2sT}}{s} + f(3T)\frac{e^{-2sT} - e^{-3sT}}{s} + \cdots$$

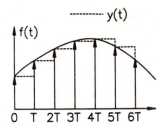

**Fig. 4.4** Zero-order hold.

$$= \frac{1-e^{-sT}}{s}[f(0)+f(1)e^{-sT}+f(2)e^{-2sT}+\cdots]$$

$$= \frac{1-e^{-sT}}{s}F(s) \tag{4.12}$$

This result shows that the Laplace transformation for the zero-order hold is

$$\frac{1-e^{-sT}}{s} \tag{4.13}$$

The characteristic of the digital controller is $D(x)$. The $Z$ transformation for the output is

$$X(Z) = \frac{G(z)D(z)R(z)}{1+G(z)D(z)} \tag{4.14}$$

Solving for the $z$ transform of the controller gives

$$D(Z) = \frac{X(z)}{G(z)[R(z)-X(z)]} \tag{4.15}$$

This result shows that the controller characteristics $D(z)$ may be obtained by knowing the plant and hold characteristics $G(z)$ and the desired response $C(z)$ to a given input $R(z)$.

### 4.2 MOLTEN POOL WIDTH CONTROL

A molten pool width $x$ can be approximated by the first-order lag state equation in terms of an input electric current $u$ [2]:

$$\frac{dx(t)}{dt} = -\lambda x(t)+bu(t) \tag{4.16}$$

where the time constant $1/\lambda$ is specified as $1.0\,s$ while the constant $b$ is specified as $0.065\,mm/AS$ in mild steel plate of $3.2\,mm$ thickness in pulse MIG welding.

The digital controller in Figure 4.3 is used to control the width of the molten pool. From Equation (4.16), the transfer function $G_p(s)$ of the plant is expressed by $b/(s+\lambda)$. The characteristic $D(z)$ of the digital controller will now be determined such that a desirable response of the molten pool width $x(t)$ to the unit step input $y(t)$ will be

$$c(t) = U_s(t-T) - e - \alpha(t-T)$$

The transfer function $G(s)$ of the plant and the zero-order hold with delay time of one period can be expressed by

$$G(s) = \frac{X(s)}{U(s)} = e^{-sT} \frac{1-e^{-sT}}{s} \frac{b}{s+\lambda} \tag{4.17}$$

The Z transform of the transfer function $G(s)$ is

$$G(z) = \frac{q}{z(z-p)} = \frac{qz^{-1}}{z-p} \qquad p = e^{-\lambda T}, \ q = \frac{b}{\lambda}(1-e^{-\lambda T}) \tag{4.18}$$

The transfer function, if arranged based on equations (4.9) and (4.10), will become the difference equation

$$x(n+1) = px(n) + qu(n-1) \tag{4.19}$$

The Z transformations for the molten pool width response

$$X(s) = \frac{e^{-sT}}{s} - \frac{e^{-sT}}{s+\alpha}$$

and for the given input

$$R(s) = 1/s$$

are expressed by

$$X(z) = \frac{1}{z-1} - \frac{1}{z-e^{-\alpha T}}$$

and

$$R(z) = \frac{z}{z-1}$$

respectively. Substituting these Z transformations in Equation (4.15) gives the characteristic $D(z)$ of the digital controller:

$$D(z) = \frac{U(z)}{E(z)} = \frac{a_0 + a_1 z^{-1}}{1 + b_1 z^{-1} + b_2 z^{-2}} \tag{4.20}$$

where

$$a_0 = \tfrac{1}{q}(1-e^{-\alpha T}), \qquad a_1 = -pa_0$$
$$b_1 = -e^{-\alpha T}, \qquad b_2 = -(1-e^{-\alpha T}) \tag{4.21}$$

or

$$U(z) + b_1 z^{-1} U(z) + b_2 z^{-2} U(z) = a_0 E(z) + a_1 z^{-1} E(z) \qquad (4.22)$$

The discrete form of this $Z$ transform yields the desired controller characteristic:

$$u(n) = a_0 e(n) + a_1 e(n-1) - b_1 u(n-1) - b_2 u(n-2) \qquad (4.23)$$

The molten pool width response for $\alpha = 1.5$ was simulated. In order to investigate the control performance when the system parameters change, the time constant $1/\lambda$ of the plant was varied, which corresponds to heat conduction. Figure 4.5 shows the simulation result with constant welding current while Figure 4.6 shows the controlled simulation result. The weld pool width is kept constant regardless of the variation of heat conduction.

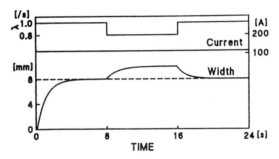

**Fig. 4.5** Simulation of weld pool width with constant current.

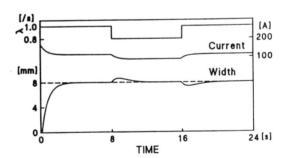

**Fig. 4.6** Simulation of weld pool width by modern control theory.

## 4.3   ADAPTIVE CONTROL

### 4.3.1   Molten pool control based on model reference adaptive control (MRAC)

The controlled object (plant) is described by

$$x[i+1] = px[i] + qu[i] \qquad (4.24)$$

where $u[i]$ and $x[i]$ are the input and output of the plant respectively. The reference model is given by

$$x_\mathrm{M}[i+1] = p_\mathrm{M} x_\mathrm{M}[i] + q_\mathrm{M} u_\mathrm{M}[i] \qquad (4.25)$$

where $u_\mathrm{M}[i]$ and $x_\mathrm{M}[i]$ are the input and output of the model respectively. The follow-up error $e[i]$ is defined by

$$e[i] = x[i] - x_\mathrm{M}[i] \qquad (4.26)$$

The objective of adaptive control is to produce the plant input $u[i]$ which makes $e[i]$ converge to zero as soon as possible when the value $i$ is sufficiently large. The following equation can be obtained from the above equation:

$$e[i+1] = p_\mathrm{M} e[i] + \underline{(p - p_\mathrm{M})x[i] + qu[i] - q_\mathrm{M} u_\mathrm{M}[i]} \qquad (4.27)$$

If it is possible to bring the underlined part of Equation (4.27) to zero, the equation can be expressed as follows:

$$e[i+1] = p_\mathrm{M} e[i] \qquad (4.28)$$

Since $|p_\mathrm{M}|$ is smaller than 1, the following equation can be obtained:

$$\lim_{i \to \infty} e[i] = 0 \qquad (4.29)$$

Bringing the underlined section of Equation (4.27) to zero gives

$$u[i] = \frac{(p_\mathrm{M} - p)x[i] + q_\mathrm{M} u_\mathrm{M}[i]}{q} \qquad (4.30)$$

Since the parameters $p$ and $q$ are unknown, they are replaced by their respective predicted quantities $\bar{p}(k)$ and $\bar{q}(k)$, and Equation (4.30) is rewritten as

$$\bar{u}[i] = \frac{(p_\mathrm{M} - \bar{p}[i])x[i] + q_\mathrm{M} u_\mathrm{M}[i]}{\bar{q}[i]} \qquad (4.31)$$

From the above equations, the follow-up error equation is redefined as

$$e[i+1] = p_\mathrm{M}[i] + (p - \bar{p}[i])x[i] + (q - \bar{q}[i])\bar{u}[i] \qquad (4.32)$$

Equation (4.32) is non-linear. In order to prove the stability of a non-linear system, Povov's hyperstabilization theory is generally adopted. Since the general MRAC theory based on Popov's hyperstabilization theory cannot be applied directly to the plant–model error expression given by Equation (4.32), the following auxiliary error signal is introduced:

$$e^*[i+1] = e[i+1] + e_\mathrm{a}[i+1]$$
$$e_\mathrm{a}[i+1] = p_M e_\mathrm{a}[i] + (\bar{p}[i] - \bar{p}[i+1])x[i] + (\bar{q}[i] - \bar{q}[i+1])\bar{u}[i] \qquad (4.33)$$

The adaptive error $e^*[i+1]$ is then given by

$$e^*[i+1] = \frac{1}{1-p_M z^{-1}} \{(p + \bar{p}[i+1])x(i) + (q - \bar{q}[i+1])\bar{u}(i)\} \qquad (4.34)$$

The well-known adaptive algorithm of the following equation based on the integration of the product of the adaptive error $e^*$ and the input and output can be used to predict $\bar{p}(i)$ and $\bar{q}(i)$:

$$\bar{p}[i] = \bar{p}[i-1] + K_p x[i-1]e^*[i] \qquad (4.35)$$

$$\bar{q}[i] = \bar{q}[i-1] + K_q \bar{u}[i-1]e^*[i] \qquad (4.36)$$

According to the general theory of MRAC, the above adaptive algorithm ensures

$$\lim_{i \to \infty} e^*[i] = 0 \qquad \text{and} \qquad \lim_{i \to \infty} e[i] = 0 \qquad (4.37)$$

The object of adaptive control is thus satisfied. The adaptive error $e^*(k)$ is obtained as follows from Equations (4.24), (4.31), (4.32), (4.33), (4.34), (4.36) and (4.37):

$$e^*[i] = \frac{p_M e^*[i-1] + x[i] - (\bar{p}[i-1]x[i-1] + \bar{q}[i-1]\bar{u}[i-1])}{1 + K_p x^2[i-1] + K_p \bar{u}^2[i-1]} \qquad (4.38)$$

Since the adaptive error $e^*$ at time $i$ must be determined by the application of $\bar{p}$ and $\bar{q}$ at time $(i-1)$ in Equation (4.38), the auxiliary signal $e_a[i]$ is introduced.

The adaptive algorithm is summarized as follows. Calculate the adaptive error $e^*[i]$ from Equation (4.38); predict $p(k)$ and $q[k]$ from Equations (4.35) and (4.36). Then, from Equation (4.31), generate the input $u$ of the plant $\bar{u}$ which allows the output of the plant to follow the output of the model. Figure 4.7 shows the block diagram of the reference model adaptive control system. It is possible to follow up with the response of the reference model

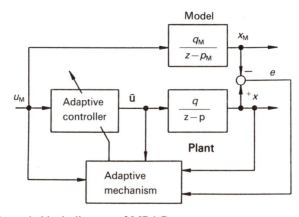

**Fig. 4.7** Schematic block diagram of MRAC.

because the parameters such as the thickness of a base metal and heat conduction can be predicted even though they are unknown or may change.

### Experimental examples

An experiment was performed to find the melting pool response when the thickness of the base metal was changed from 4.5 mm to 3.5 mm over a length of 100 mm. Figure 4.8 shows the result of molten pool width without the reference adaptive controller. Figure 4.9 shows the adaptively controlled welding current and weld pool width.

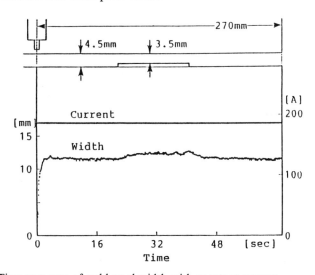

**Fig. 4.8** Time response of weld pool width with constant current.

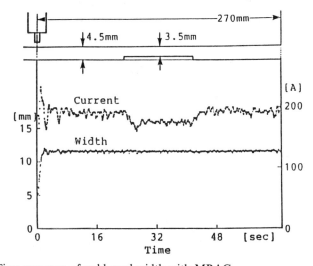

**Fig. 4.9** Time response of weld pool width with MRAC.

The reference value of the molten pool width was 11.2 mm, and the model was set with the time constant 0.43 s. In Figure 4.9, the gains $K_p$ and $K_q$ for estimating the parameters, are defined as below:

$$K_p = 8.5 \times 10^{-5} \text{ mm}, \qquad K_q = 3.8 \times 10^{-6} \text{ A}^2 \qquad (4.39)$$

### 4.3.2  On-line optimization based on calculated model for welding conditions

Attempts have been made to use a numerical calculation model as a basis for an optimization method to maintain a required penetration shape. This is a method of identifying parameters from the temperature information at welds [5] and controlling weld pool shape by adjusting the welding current.

Consider the simple case of TIG welding bead on plate. Figure 4.10 shows the calculation model. As illustrated in Figure 4.10, the arc is assumed to be a uniformly distributed heat source. The calculation mainly comprises the difference method for a semi-steady-state heat conduction equation.

Figure 4.11 shows the principle of temperature measurement. An infrared sensor was used to measure the temperature of the welds. The arc current

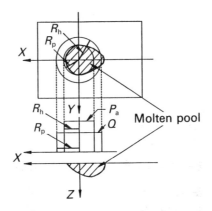

**Fig. 4.10** Computation model: $Q$, heat input; $P_a$, arc pressure.

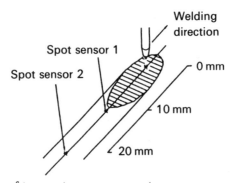

**Fig. 4.11** Principle of temperature measurement.

should be reduced during measurement to avoid the adverse effect of the arc light. Figure 4.12 shows the principle control system. The required bead dimensions (such as penetration depth $D_0$) are given as required values, and then the welding parameters (current etc.) required to maintain the required value are calculated using the model. This method uses a computer-based model to optimize the following objective functions and determine the required welding parameters:

$$F=(D_C-D_0)^2\rightarrow\min \qquad (4.40)$$

where $D_C$ is the value calculated from the model, and $D_0$ is the required value.

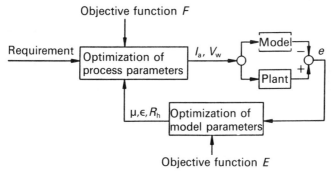

**Fig. 4.12** Control system.

In determining process parameters, the model parameters such as heat efficiency $\mu$ and heat source $R_H$ are optimized in turn. This may be in response to temperature measurement as shown in Figure 4.11, and by optimizing the objective function $E$ below:

$$E=(T_1-T_{S1})^2+(T_2-T_{S2})^2\rightarrow\min \qquad (4.41)$$

where $T_1$, $T_2$ are model-based calculated values, and $T_{S1}$, $T_{S2}$ are measured values. This operation should be repeated periodically.

*Experimental results*

Figure 4.13 shows one example of the experimental results. In this example, the penetration depth ($D_0 = 3$ mm) was given as a required value. However, the heat efficiency $\mu$ was adopted as the model parameter which could be identified. Figure 4.13 shows the changes in current value $I_a$, measured temperature, and penetration depth $D_t$, which was obtained from the bead cross-sectional shape after welding. The experiment shows that the welding current is substantially stabilized after the second revision, while the penetration depth approaches the required value as shown in the figure. It was also confirmed that the experimental results were virtually satisfactory in the back bead welding of sheets.

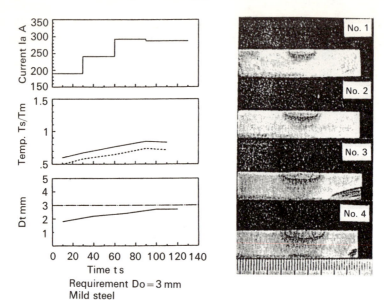

**Fig. 4.13** Experimental results.

## 4.4   FUZZY CONTROL

### 4.4.1   Fuzzy inference method

Fuzzy theory is a concept based on the set theory proposed by Professor Zadeh of California University. Figure 4.14 shows the outline of a fuzzy set in terms of the example of molten pool width $W$. The membership grade $P$ of $W$ in the crisp set $P$ (ordinary set) has two values only, 1 and 0. If $P=0$ then $W$ does not belong to $P$. If $P=1$ then $W$ belongs to $P$. For example, when a reference value is assumed as 9.5 mm, the width which exceeds 10.7 mm is defined as 'wider'. When the width is 10.6 mm, it can be clearly classified as not wide. In contrast with this, the fuzzy set is based on a more subjective conception, such as 'wide' and 'narrow'. The boundary is not clearly defined. The membership grade of $W$ in the fuzzy sets 'wide $P$', 'medium ZO' and 'narrow N' are expressed by a membership function $h(W)$, which has a value ranging from 0.0 to 1.0 as illustrated in Figure 4.14 (a).

Fuzzy inference refers to the inference method which introduces one conclusion from a number of fuzzy rules. Consider the following molten pool width control as an example:

*Rule 1:*     If the molten pool is narrow in width and the width is hardly increasing, then increase the electric current.

*Rule 2:*     If the width of molten pool is correct and the width is hardly increasing, then maintain the electric current.

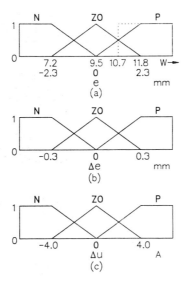

**Fig. 4.14** Trigonal fuzzy variable.

*Fact*:          The width of molten pool is rather narrow and there is almost no increase in the width.

*Conclusion*:   Increase the electric current slightly.

For practical application, it is necessary to obtain the fuzzy set of 'increase the electric current slightly'. Several types of fuzzy inference methods are available. The most popular one is adopted here. Figure 4.14 shows the fuzzy variables (membership functions) by using the basic three triangles where N stands for 'negative', ZO for 'zero' and P for 'positive'.

In Figure 4.14, '$e$' is the deviation ($e_n = x_n - r$) of a molten pool width $x_n$ from the reference value $r$, while $\Delta e$ is the variation ($\Delta e_n = e_n - e_{n-1}$). In the next section they correspond to $x_1$ and $x_2$ respectively. $\Delta u$ is a variable variation of current, which corresponds to $y$. The control rule R is represented as follows:

(Condition part)          (Operation part)

R1 If $x_1$ is $A_{11}$ ($=$P) and $x_2$ is $A_{12}$ ($=$ZO) then $y$ is $B_1$ ($=$P)

R2 If $x_1$ is $A_{21}$($=$ZO) and $x_2$ is $A_{22}$ ($=$ZO) the $y$ is $B_2$ ($=$ZO)

Fact          $x_1$ is $x_1^0$, $x_2$ is $x_2^0$

Conclusion          $y$ is $B_0$

where $A_{11}$, $A_{12}$, $A_{21}$, $A_{22}$, $B_1$ and $B_2$ stand for the fuzzy sets of P, ZO and N.

Fuzzy inference is illustrated in Figure 4.15. Let the actual value of $e$ and of $\Delta e$ be $e^0 = -1.7$ mm and $\Delta e^0 = 0.2$ mm, respectively. At first an attempt must be made to find how far $x_1$ and $x_2$ conform to the condition of the if-part in each rule and to determine the degree of conformity. Let the fuzzy sets of $x_1$ and $x_2$ in the $i$th rule be $A_{i1}$ and $A_{i2}$ respectively.

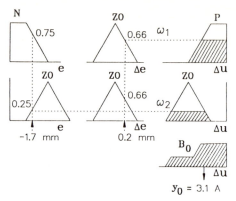

**Fig. 4.15** Fuzzy inference.

The degree of conformity $\omega_i$ can be determined by the following equation:

$$\omega_i = h_{A_{i1}}(x_1^0) \wedge h_{A_{i2}}(x_2^0) \tag{4.42}$$

where the calculation of $a \wedge b$ produces the minimum of $a$ and $b$. This means that a lower-grade value is adopted when there are two conditions in the if-part. A value of 0.66 is adopted in the if-part in rule 1 while 0.25 is adopted in rule 2.

The inference result $B_i^*$ for the $i$th rule is represented by the following equation:

$$B_i^*(y) = \omega_i \wedge B_i(y) \tag{4.43}$$

This equation means that the fuzzy set $B_i$ is cut with degree of conformity $\omega_i$. The output of the operation part is reduced with the degree of conformity. Figure 4.15 shows the inference result for two rules where results in the portions higher than 0.66 and 0.25 have been cut away from fuzzy sets $B_1$ and $B_2$ respectively. The result $B_0$ of the whole inference is represented by the following equation:

$$B_0 = B_1^* \bigcup B_2^* \tag{4.44}$$

where the result $B_0$ is expressed by the union of two fuzzy sets (Figure 4.15). It is impossible to output the fuzzy set $B_0$ as a manipulable variable. From the resulting fuzzy set $B_0$ we must determine the concrete value $Y$. The manipulable variable $Y_0$ is determined by means of a centre of gravity under the membership function $h_{B_0}(y)$ in the fuzzy set $B_0$ (Figure 4.15: $y_0 = \Delta u = 3.1$ A).

The methods for deciding the control rules (Table 4.1) and fuzzy variables comprise:

1. methods based on the experience and knowledge of experts;
2. adjustment of the control rules and fuzzy variables with the plant models;
3. fuzzy models of the plant.

**Table 4.1** Control rules

| $\Delta e$\\$^e$ | N | ZO | P |
|---|---|---|---|
| P | ZO | N | N |
| ZO | P | ZO | N |
| N | P | P | ZO |

P = positive; ZO = zero; N = negative

### 4.4.2 Molten pool control based on fuzzy inference

Figure 4.16 shows a molten pool width control system using fuzzy inference. According to the control rules, this system infers a variation $\Delta u$ of a manipulable variable $u$ (input current) from the deviation $e \ (= x - r)$ from a reference value $r$ of the molten pool width and the variation $\Delta e$ of the deviation $e$. The simulation is carried out in the case where the heat conduction $\lambda$ changes 8 s after the start.

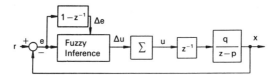

**Fig. 4.16** Fuzzy control system.

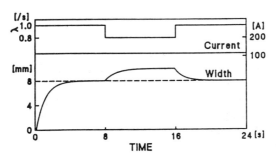

**Fig. 4.17** Simulation of weld pool width in constant current.

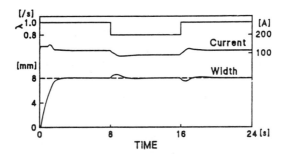

**Fig. 4.18** Simulation of weld pool width with fuzzy controller.

*Basis for welding process control*

First, a constant current (120 A) is applied. The result is shown in Figure 4.17, where the width changes with the variation of heat. Next, the simulation is performed with the fuzzy controller. The result is shown in Figure 4.18. The width is kept constant regardless of the variation of the heat conduction.

## REFERENCES

1. Satake, I. (1977) *Linear Algebra*, Syoukabou Publishing Co. Ltd.
2. Ohshima, K., Morita, M., Fujii, K., Yamamoto, M. and Kubota, T. (1987) Observation and digital control of weld pool in pulsed MIG welding. *Journal of the Japan Welding Society*, **5** (3), 18–25.
3. Mita, T., Hara, S. and Kondou, R. (1988) *Introduction to Digital Control*, Corona Publishing Co. Ltd.
4. Yamamoto, H., Kaneko, Y., Ohshima, K. *et al.* (1988) Adaptive control of pulsed MIG welding using image processing system, in *Conf. Rec. of the IEEE IAS*.
5. Nishiguchi, K., Ohji, T. and Hiramoto, S. (1987) An on-line optimization of welding parameters by a numerical model, in *Proceedings Japan Welding Society 40th Convention*.
6. Sugeno, M. *et al.* (1985) *Industrial Applications of Fuzzy Control*, North-Holland.

# CHAPTER 5

# Sensor techniques and welding robots

The arc welding robot is designed to repeat a specified job, and is programmed to weld sequentially specified parts on an assembly. However, it is difficult for such robots to trace a welding line on an assembly or compensate for a change in the geometry of a groove, and thus obtain high-quality welding.

The welding robot must therefore be equipped with a sensor designed to trace any change produced on an assembly in order to maintain high-quality welding performance. This may require an adequate level of artificial intelligence (AI) and the ability to analyse the information obtained from the sensor. More specifically, the required functions must decide the optimum welding condition, and the position of the tip of a welding torch in relation to the welding line based on the information obtained from the sensor, must instruct a robot to operate based on AI in real time, and must change the predetermined sequence of the work. Such an autonomous robot is classified as an intelligent robot.

Figure 5.1 shows the block diagram of an autonomous welding robot. As illustrated, the robot comprises a closed-loop control system in a three-layer structure. The first layer control of position, posture and manipulation of the welding torch is a control of each joint of a robot manipulator. Specifically, each joint is equipped with a position feedback device, such as an encoder. The second layer is provided with a sensor to detect the state of welding. The sensor communicates via an interface and an intelligent algorithm. The control loop is closed by an interpolation algorithm. This changes and modifies the original work program based on the revised information obtained from the AI algorithm. The third layer is a control loop for the welding parameters (arc current, voltage and waveform). Like the second-layer loop, this control loop is designed to revise the originally programmed welding condition based on information obtained from the artificial intelligence algorithm.

*Sensors and Control Systems in Arc Welding.* Edited by Hirokazu Nomura.
Published in 1994 by Chapman & Hall, London. ISBN 0 412 47490 5

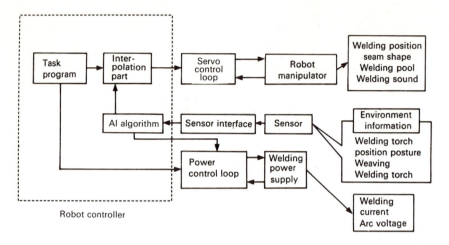

**Fig. 5.1** Sensor system for arc welding robots.

## 5.1   PRESENT STATUS OF ROBOT SENSORS

Further research and development effort is required both on sensors and on the related technologies, to realize the type of system illustrated in Figure 5.1 and hence allow welding robots to perform a really intelligent role.

In general, sensors for arc welding robots must meet the following requirements.

1. They must be compact and lightweight, thereby not interrupting welding work.
2. They must be highly reliable and serviceable in real time since they are to be used on an automated production line.
3. They must be highly resistant to adverse welding environments, such as welding heat, fumes, and sputtering.
4. They must be highly resistant to noise in terms of disturbances.
5. They must be low cost.

The sensors currently available, as described elsewhere in this book, are based on the application of electromagnetism, optics, ultrasonics, and welding current and voltage, in various practical configurations. The current status of robot engineering still demands further progress in the field of artificial intelligence in order to allow a welding robot to operate autonomously. It will probably be some time before welding robots are capable of carrying out fully automatic operations. This is clear if human functions and robot functions are compared. No robot can compete with the superiority of human senses and intelligence.

It is therefore essential for humans to assist the currently available sensors by setting and selecting the relevant parameters, such as motion, sequence and condition.

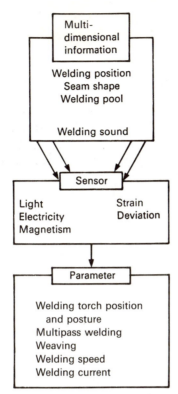

**Fig. 5.2** Multidimensional information requirements for an arc welding robot.

Robot users demand the multidimensional information indicated in Figure 5.2. For arc welding, the sensor system needs to utilize information about welding positions, joint geometry, weld shapes, welding pool state, and arc state to vary the position of welding torch, welding posture, welding speed, arc current and arc voltage. This is the knowledge and skill acquired naturally by the manual welder during his work experience.

The sensors currently available on the market are unidimensional and exclusively limited to special applications to detect specific objects under specific environments. Error detection often takes place. Even so-called 'intelligent seniors' are of such a low level that their low accuracy and low reliability must be backed up with software for practical application. Therefore, the users cannot obtain the required parameters and find this unsatisfactory.

To perform satisfactory autonomous robot arc welding requires selection of the optimum sensor. It also requires judgement, to extract the requisite information from an approximate arrangement of the sensors. Recent papers have described an intelligent robot designed to control the depth of penetration and bead simultaneously based on the combined application of an arc sensor and an image-processing sensor [1], and an adaptive control system which uses a CCD camera and a rotating-mirror laser displacement sensor [2]. The use of such sensor systems, based on the arrangement of different sensors, is expected to increase in the future.

## 5.2  CONNECTION BETWEEN SENSOR AND ROBOT

Two levels of interface are necessary between a robot controller and a sensor, as shown in Figure 5.3. In interface (a), the information detected by a sensor is converted into a signal wavefom, the features of which determine the specification and design of the interface. In interface (b), a digital signal serves as the interface. Currently, both interfaces (a) and (b) are dependent on the performance of a sensor, and the interfacing method varies with the selected sensor. So even when an attempt is made to select a sensor to obtain the required information, the sensor itself is limited to a specific application. This is one of the factors which prevents sensors from being widely used.

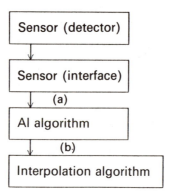

**Fig. 5.3** Construction of sensor.

The signal configuration of (a) may be on/off, analogue, or digital. It varies with the detection level of a sensor and its type. Digital transmission forms the basis of interface (b), and may be bus direct connection, point-to-point Centronics, RS-232C, RS-422/423/485, or RS-488 (GPIB). However, physical interface protocols are required to perform digital transmission and no protocols are available for sensor information. It is therefore necessary to change the way the information is sent for each type of sensor.

An example of the interface between a laser scanning sensor and a robot is given in [4].

Figure 5.4 shows an example of an added communication subset with a control subset installed to connect a sensor to a robot.

The communication subset is designed to integrate the communication protocols for the robot controller and the control subset and comprises a microprocessor, operating under multitasking control in real time.

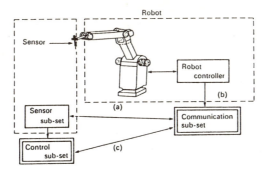

**Fig. 5.4** Connection of sensor to robot.

Since a robot has no coordinate conversion function, the control subset includes a mechanism model of a robot manipulator, thereby converting various kinds of coordinates into programs which allow the robot to operate.

The control subset is capable of adjusting the offset between the mounting positions of a sensor preceding a tool so that it can transmit the information from the sensor accurately, and convert the coordinates of the sensor into the joint coordinates of the robot.

The required interface element involves high-speed performance and enhanced multifunctions. When attempting to construct a sensor-based high-intelligent robot, it is very important to speed up the transmission rate and expand the functions. Such features are increased if multiple sensors are used to obtain required parameters, and so it is necessary to establish flexible standardization of interface (b).

## 5.3   ROBOT OFF-LINE PROGRAMMING

The programming methods by which the trajectory and functions of an arc welding robot are created are currently dominated by the 'teaching' method. This requires the utilization of working robots in service, object work, and production line facilities such as peripheral devices, in order to determine the operating procedures of an arc welding robot and to quantify the parameters of the welding condition. The welding robot plays an essential part in the course of programming. This method is easy to apply, and allows a programmer to grasp the actual working situation. Program proving must also be carried out using the production facility. However, the recent tendency towards low-volume production of a wide variety of products calls for more flexible production systems. This creates problems. It degrades the productivity of robots on the production line in the course of programming, and hinders the preparation of advanced programs, such as the incorporation of conditional branching into the programs.

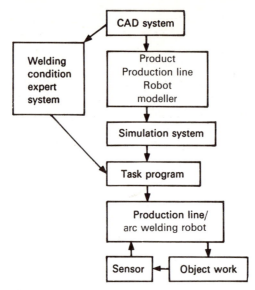

**Fig. 5.5** Conceptual robot architecture.

Recently, much research and development effort has been devoted to the introduction of preproduction systems for the preparation of working programs for robots. This is called *off-line programming*. It makes it possible to describe a working program, simulate the application of production line tools, and execute programs including the preparation of welding conditions without using any robot in service. The core of off-line programming is the robot control language, which may be different for various types and makes of robot. The robot language or the off-line programming system is evaluated by whether or not it allows an operator to write, easily and freely, an operation program to drive a welding robot.

Also, the welding robot program must incorporate welding conditions or process parameters. An off-line programming system must therefore be capable of acquiring these. Typically, the programmer will enter details of the work, the joint to be made, the material and thickness, and the joint groove geometry. The programming system will select appropriate parameters from a welding procedure database. An expert system may be used to select the optimum conditions from those available.

Such advances in programming languages and process condition selection systems have reduced the need for programming of the production system on line. Off-line programming systems may also offer higher functions such as graphical simulation, which enables the proving of program execution, also off line, and can be used too for such tasks as developing optimum welding cell layouts, and assisting in robot selection.

Figure 5.5 is a conceptual diagram of architecture for robots and robot sensors installed on the production lines which include off-line programming.

Accompanied by the marked progress in computer-aided design (CAD) and computer-aided manufacturing (CAM), even arc welding is expected to develop to CAM which delivers a work program directly to a robot on the production line based on a drawing designed by CAD.

As described above, the future arc welding robot will be organically coupled with off-line programming, robot sensors, and robot controllers.

Figure 5.5 shows the robot architecture of the future. As for robot sensors, there are many problems, awaiting solutions. It is therefore likely to be some time before the ideal robot architecture comes into existence.

## REFERENCES

1. Sugitani, Y., Nishi, Y. and Sato, T. (Chapter 44 in this book).
2. Fujita, K. (Chapter 18 in this book).
3. Committee of Research for Robotics End Effector (1987) Report of research for robotics end effector. Robotics Society of Japan.
4. Clocksin, W.R., Bromley, J.S.E. and Davey, P.G. (1985) An implementation of mode-based visual feedback for robot arc welding of thin sheet steel. *International Journal of Robotics Research*, **4** (1), 13–26.

# CHAPTER 6

# Analysis of questionnaire results on sensor applications to welding processes

The Editorial Committee for this book has investigated current applications of arc welding sensors and control systems, and the problems that remain to be solved using a questionnaire (set out below) similar to the one distributed in 1981 by the Technical Commission on Welding Processes [1].

1.  Please enter the name of the sensor, chosen from the attached list. If it is not listed, please describe its construction and function.
2.  State category of sensor:
    (1)  By application
        A.  Seam tracking (including torch height control)
        B.  Welding process control
        C.  Both A and B
        D.  Monitoring
        E.  Other (please specify)
    (2)  By operating principle
        A.  Arc phenomena
        B.  Electromagnetic
        C.  Optical
        D.  Temperature and heat
        E.  Acoustic emission and ultrasonic
        F.  Fluidics
        G.  Contact
        H.  Other (please specify)
3.  Describe the detection principle of the sensor and illustrate the control system based on it. Ensure that you include the detection object of the sensor and the control object.
4.  Were the sensor and control system purchased, or developed in your company? When? Please give details.

*Sensors and Control Systems in Arc Welding.* Edited by Hirokazu Nomura.
Published in 1994 by Chapman & Hall, London. ISBN 0 412 47490 5

5. Has the sensor been used on the production line, or marketed? When? Please give details.
6. If the answer to question 5 is YES, please state the number of sensors and control systems, the application, and when the system became operational.
7. Is the application of the sensor and control system limited in scope? If so, please specify:
   (1) Welding method (e.g. $CO_2$, MIG, TIG, SAW)
   (2) Welding conditions (welding current, voltage welding speed)
   (3) Welding materials (e.g. type of wire, size, shielding gas)
   (4) Base metal (e.g. carbon steel, stainless steel)
   (5) Thickness of plate and groove shape
   (6) Welding position
   (7) Other (please specify)
8. Please list the good points and advantages of the sensor and control system:
   (1) Improved quality and stabilization
   (2) Compensation for material positioning and faulty material processing accuracy
   (3) Labour saving and compensation for the shortage of man power
   (4) Compensation for reduced operator skill levels
   (5) Improved working environment
   (6) Improved productivity and efficiency
   (7) Other (please specify)
9. Please list the problems and disadvantages of the sensor and control system:
   (1) Accuracy, repeatability, and responsiveness
   (2) Spatter resistance
   (3) Arc light resistance
   (4) Fume resistance
   (5) Noise resistance
   (6) Heat resistance
   (7) Durability
   (8) Shape and size
   (9) Difficulty of maintenance
   (10) Difficult to design and manufacture hardware and software
   (11) Price
   (12) Other (please specify)
10. Will you need to improve the sensor and control system in the future? If so, please give details.
11. Please comment on the future of the sensor and control system:
    (1) The application scope will be extended.
    (2) The application scope will be extended once the system has been improved.
    (3) The current situation will be maintained.

(4) The operation will be scaled down or suspended

(5) Other (please specify)

12. What are your views on the future of sensors and control systems?

In the questionnaire, the sensor and control system were defined as follows: 'In automatic arc welding, when external and internal changes which affect welding results are captured by a detector, and when monitoring, operation and control are effected by this detected signal, the detector is regarded as a sensor and the whole control device is defined as a sensor system (control system).' Table 1.1 in Chapter 1 shows the applications of sensors that fall within the scope of the survey, and those that were excluded. The questionnaires were distributed to 143 members of the Technical Commission on Welding Processes and the Technical Committee of Welding Arc Physics (both in the Japan Welding Society) and Robot Welding (JWES). The number of replies received was 92 (a 64.3% response rate).

Figure 6.1 shows the distribution by industries of welding sensors and control systems, based on the replies received.

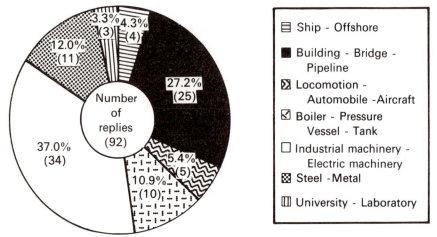

**Fig. 6.1** Industrial use of welding sensors.

Figure 6.2 shows the applications of sensors, and contrasts the results with those of the previous survey. In 1981, seam tracking accounted for 80% of sensor use. Although this percentage has dropped slightly, sensors used exclusively for seam tracking still account for 70% of applications, and exceed all other types of sensor. Sensors for welding process control, or continued seam tracking control and welding process control, have increased by 10% compared with the previous investigation, but their application is still less than 30% in all. Applications to the automation of arc welding are still at an early stage of development.

Figure 6.3 shows the classification of sensors in service by method of operation. They mainly comprise arc sensors (33.7%), optical sensors

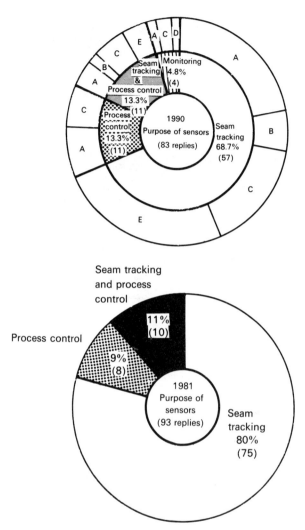

**Fig. 6.2** (a) Sensor applications as reported in 1990 survey: A, arc phenomena; B, electromagnetic; C, optical; D, temperature; E, contact. (b) Results of 1981 survey.

(28.9%) and contact sensors (28.9%). Self-developed arc sensors and optical sensors outnumber purchases, and the sales ratio of self-developed sensors has increased. However, contact sensors are mostly purchased. In 1981, contact sensors (electrical contact probe, potentiometer, DTF, wire contact and combined types) exceeded other types. It is clear that non-contact types such as arc sensors and optical sensors have made remarkable progress, greatly outnumbering contact sensors. The applications of sensors have not changed noticeably, but the types of sensor in service are much more advanced and sophisticated ones, revealing a move towards more advanced automatic control in the future.

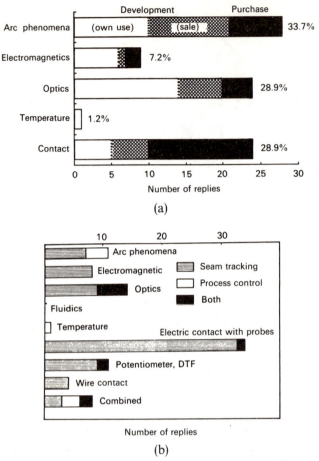

**Fig. 6.3** Classification of sensors by method of operation: (a) 1990 survey; (b) 1981 survey

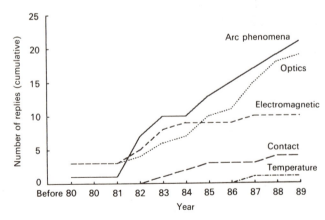

**Fig. 6.4** Growth in self-developed sensors, 1981–1990.

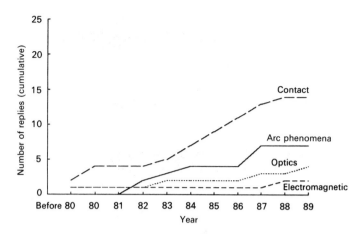

**Fig. 6.5** Growth in purchased sensors, 1981–1990.

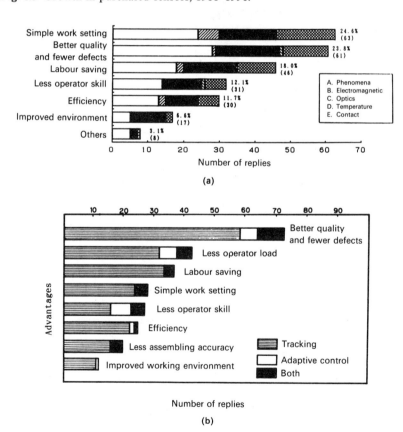

**Fig. 6.6** Advantages of sensors: (a) 1990 survey; (b) 1981 survey.

Figures 6.4 and 6.5 show the growth of self-developed and purchased sensors since the previous survey. It is clear that arc sensors and optical sensors have been continuously developed at a dramatic rate since 1981. Electromagnetic sensors were developed at almost the same rate to begin with, but they have been virtually static since 1984. With regards to purchased sensors, the contact types have dominated the others consistently. They have increased markedly since 1982, but appear to have reached saturation since 1988.

The growth in sales of arc sensors and optical sensors is not dramatic, but their steady increase shows the first stages of practical application. Optical

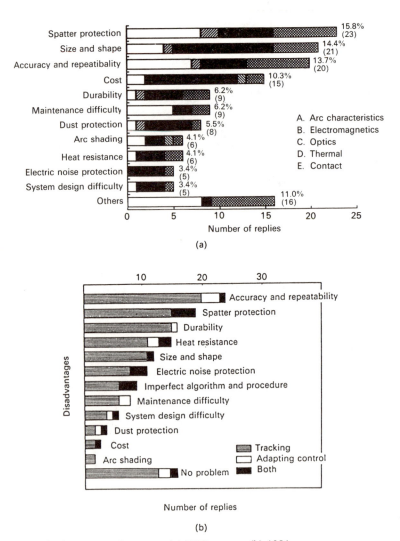

Fig. 6.7 Disadvantages of sensors: (a) 1990 survey; (b) 1981 survey.

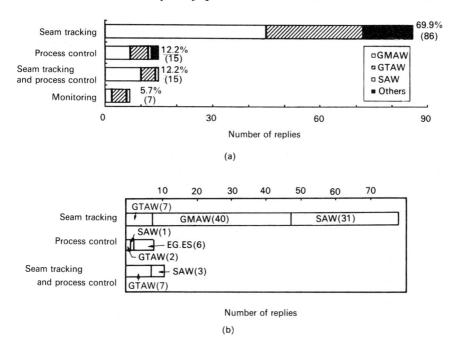

**Fig. 6.8** Welding process applications of sensors: (a) 1990 survey; (b) 1981 survey.

sensors have started a steady increase, but their use is still at a low level. Arc sensors and optical sensors, which show marked progress in Figure 6.4, have entered the stage of practical application and wide use, thus creating a market value.

Figure 6.6(a) shows the reasons for introducing a sensor system or the perceived advantages of sensors already in service. When these results are compared with the 1981 analysis, the impression emerges that companies are constantly trying to maintain the desired product quality and improve productivity, despite suffering from a chronic shortage of skilled workers.

Figure 6.7(a) indicates the problems and disadvantages of the arc welding sensors currently available. The main problems are inadequate protection against spatter, larger shape, and limitations of accuracy, repeatability, and responsiveness. High cost also appears to be a problem. Other problems relate mainly to the question of applicability. For example, for arc sensors, the groove wall and oscillation are necessary because of the method of operation, and this imposes limitations on applications to thin sheet welding. With contact sensors, it is virtually impossible to introduce real-time sensing since the memory playback system is adopted and the efficiency is poor. Figure 6.7(b) shows the disadvantages indicated in the previous survey. Although some problems, such as protection against spatter, have not been solved yet, other problems have been largely dealt with. However, the practical problems of shape, dimensions and cost are increasingly significant.

When looking at the type of arc welding process to which sensors are applied, it is found that consumable electrode arc welding processes (such as MIG, MAG and $CO_2$) accounts for nearly 50% in nearly every application field (Figure 6.8), even though the welding phenomena involved are more sophisticated than those of non-consumable electrode processes such as TIG and plasma. This reflects the considerable amount of work that has gone into investigation of arc welding phenomena.

'Arc sensors' account for a third of all the replies received. Table 6.1 shows the survey results on the restriction of the application of arc sensors. Conventionally, the application of arc sensors to short-circuiting welding have been considered difficult, however, from the replies, it seems that there are no marked restrictions on welding current values. Moreover, in the thin plate industry, which has introduced arc welding robots to a considerable

**Table 6.1** Limitations on applications of arc sensors

| Welding parameter | Limitation | Number of replies |
|---|---|---|
| Shielding gas | Both MAG and $CO_2$ | 17 |
| | MAG only | 3 |
| | CO only | 0 |
| Wire diameter | $\leqslant 2.0$ mm | 1 |
| | $\leqslant 1.6$ mm | 7 |
| | $\leqslant 1.4$ mm | 1 |
| | Only 1.2 mm | 4 |
| Welding position | Flat | 7 |
| | Flat and vertical up | 4 |
| | All positions | 2 |
| | Flat, vertical down, overhead and horizontal | 1 |
| Plate thickness: minimum thickness for seam tracking in lap joint welding | $\geqslant 1.2$ mm | 1 |
| | $\geqslant 2.0$ mm | 2 |
| | $\geqslant 2.3$ mm | 3 |
| Welding speed | $\leqslant 0.5$ m/min | 1 |
| | $\leqslant 0.5$ m/min | 1 |
| | $\leqslant 1.0$ m/min | 2 |
| | $\leqslant 1.2$ m/min | 2 |
| | $\leqslant 1.25$ m/min | 1 |
| | $\leqslant 1.4$ m/min | 1 |
| | $\leqslant 2.5$ m/min | 1 |
| Welding current | $\geqslant 70$ A | 1 |
| | $\geqslant 80$ A | 1 |
| | $\geqslant 90$ A | 1 |
| | $\geqslant 100$ A | 2 |
| | $\leqslant 120$ A | 2 |
| | $\geqslant 140$ A | 1 |
| | $\geqslant 200$ A | 2 |

extent, more attention is focused on applications to thinner plate with high-speed welding. The arc sensors currently used comprise mostly reciprocating-type oscillation, whose application is restricted to plate thickness over 2.0 mm and welding speed below 1.0 m/min. However, by use of high-speed rotation of the arc, one example shows high performance of 1.2 mm thickness and 2.5 m/min welding speed. With regard to welding positions, the flat position is most commonly used, but it can be seen that applications to the vertical-up, vertical-down and overhead positions are being promoted.

## REFERENCES

1. Masumoto, I. *et al.* (1982) Development and application of sensors and sensor systems for arc welding in Japan. IIW Document XII-C-031-82.

CHAPTER 7

# Future trends

As described in the previous chapter, a comparison between the current and the 1981 surveys shows no marked changes in the purposes for which sensors are used, but changes are observed in the types of sensors, and their functions are found to have improved to a satisfactory extent.

Figure 7.1 shows the responses to the question about the future trends for arc welding sensors and control systems. The survey shows a remarkable change in users' opinions. In 1981 positive opinions, such as 'the application

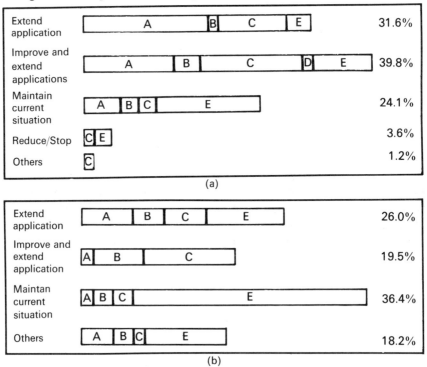

**Fig. 7.1** Future trends: (a) 1990 survey (83 replies); (b) 1981 survey (77 replies). A, arc phenomena; B, electromagnetic; C, optics; D, temperature; E, contact.

*Sensors and Control Systems in Arc Welding.* Edited by Hirokazu Nomura.
Published in 1994 by Chapman & Hall, London. ISBN 0 412 47490 5

**Table 7.1** Future expectations for arc welding sensors

| Type | Number of replies |
|---|---|
| Arc phenomena | 8 |
| Magnetic | 0 |
| Optical (laser, image processing) | 25 |
| Temperature | 1 |
| Acoustic emission, ultrasonics | 0 |
| Fluidics | 0 |
| Contact | 0 |
| Other | 18 |

**Table 7.2** Future expectations for control systems

| Application | Number of replies |
|---|---|
| High-speed welding | 1 |
| Adaptive control | 8 |
| Simple systems | 1 |
| Welding result diagnosis | 2 |
| Fuzzy control | 1 |
| Artificial intelligence | 1 |

will be extended' and 'the application will be extended if they are improved' accounted for 42% of the total. In the current survey, such positive opinions have increased to 71%, while pessimistic views have decreased dramatically from 20% to 5%.

Table 7.1 shows the results of the survey on 'the sensors which are expected to make significant progress in the future'. Nearly 50% of the respondents predicted that the use of optical sensors will expand in the future. Only 15% supported arc sensors, which are the leading method at present. Expectations for non-contact sensors, without specifying the operating method but stressing the need for low cost, small size and durability, reach 35%. Table 7.2 shows the results of the survey on 'control systems which are expected to make significant progress in the future'. The majority of respondents predict that adaptive control systems will be established for arc welding. Some predict the appearance of a control system which is able to diagnose the welding result. There is a clear need to develop, in due course, a fully automatic arc welding method.

# CHAPTER 8

# Laser sensing methods for groove tracking control

*Hirohisa Fujiyama*

Recently, laser control systems have brought rapid advances in processing and high productivity. This paper describes in detail the development of laser sensoring methods for groove tracking control.

## 8.1 INTRODUCTION

The development of control systems utilizing laser technology has been of considerable importance in advancing industrial machines and processes, resulting in significant savings in labour cost and higher productivity. In particular, the semiconductor laser, with different optical characteristics compared with He–Ne and $CO_2$ gas lasers, offers great potential for application in short-run work on a large variety of products. We have developed a weld groove tracking control system using a laser sensing method applied to three-dimensional measurement. The details of the method are described below.

## 8.2 OUTLINE OF GROOVE TRACKING CONTROL BY LASER SENSING

### 8.2.1 Laser sensing by optical beam scanning

Automatic welding systems and arc welding robots often require intelligent groove tracking by laser sensing. A schematic diagram of the method is shown in Figure 8.1. It is used for plate preparation, in which a scanning

*Sensors and Control Systems in Arc Welding.* Edited by Hirokazu Nomura.
Published in 1994 by Chapman & Hall, London. ISBN 0 412 47490 5

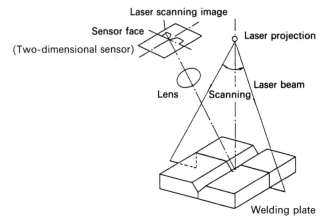

**Fig. 8.1** Principle of laser scanning for V groove detection.

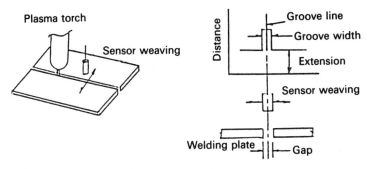

**Fig. 8.2** Laser detection of I groove in plasma welding.

laser beam spot is employed in seam and groove gap tracking. Such a system may also be used for molten metal height control. Figure 8.2 shows a schematic diagram of the method applied to the plasma welding process.

### 8.2.2 Differential distance laser sensing

This method comprises twin differential distance type laser sensors laser beam spot at an inclined position relative to a V groove joint. Groove seam-tracking is in proportion to each groove face distance inferred from the output signal, the control system being a closed feedback circuit. Groove width determination is also possible.

A schematic diagram of the method is shown in Figure 8.3, applied to the gas shielding arc welding process.

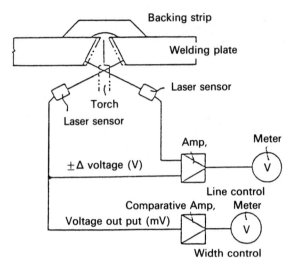

**Fig. 8.3** Laser detection of V groove in gas shielding welding.

## 8.3   CAPABILITIES AND APPLICATIONS

### 8.3.1   Laser sensing by optical beam scanning

For an open square butt joint (I groove) the laser beam spot is scanned at right angles with a laser spot diameter of 0.5–2.00 mm. It is then possible to detect a groove with minimum sensing accuracy of 0.2–0.3 mm. The required distance from laser to welding arc position is 100 mm minimum.

The control method is shown in Figure 8.4. It consists of a CPU controller to detect pulse width and to control twin motors. To keep the correct distance of the groove face to the plasma torch (extension), one of the motors is controlled by the differential signal voltage $\pm \Delta E = 0$ (V) obtained by comparing the threshold set voltage $SE_h$ with the base detecting signal voltage. A narrow groove gap requires a low filler wire feed value, and vice versa. Groove width is monitored by comparing the sensor output with a threshold set voltage $SE_f$. The measured groove width is used to select an appropriate wire feed speed from an I groove database.

### 8.3.2   Laser sensing: differential distance method

The control method is shown in Figure 8.5.

Groove seam tracking is made possible by maintaining the comparative threshold difference voltage of the two sensors $\pm \Delta E = 0$ (V) by adjustment of the torch position. Travelling speed is adjusted according to the measured groove width: slower travelling speed at wider groove width, faster travelling speed at narrower groove width. It is possible to apply this system to reverse

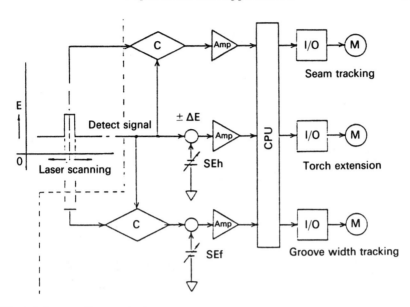

**Fig. 8.4** System diagram for laser detection of I groove in plasma welding.

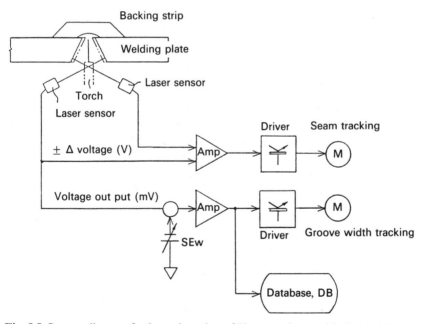

**Fig. 8.5** System diagram for laser detection of V groove in gas-shielded welding.

side welding, using the control circuit to select the correct welding conditions. The twin laser distance sensor has good anti-noise and temperature drift characteristics.

## 8.4  FEATURES AND NATURE OF DISTANCE-TYPE LASER SENSOR

Recently developed lasers are practical, economical and easy to use. Of particular importance are the newly developed optical fibre scope, optical sensor and laser oscillator and their possible applications to laser technology.

The major application features of a laser sensor include:

1. linearity of the beam;
2. ease of use;
3. a true monochromatic light beam;
4. ease of control;
5. low interference.

In comparison, conventional optical light beams require carefully developed optical lens systems. The monochromatic laser beam enables use of a narrow bandwidth filter to separate it from the optical noise of the weld-

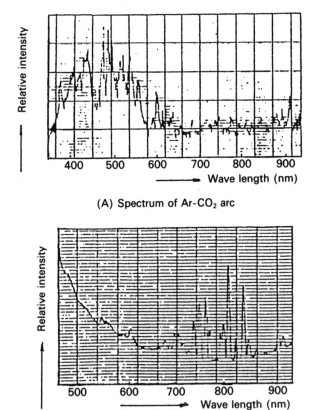

(A) Spectrum of Ar-$CO_2$ arc

(B) Spectrum of TIG arc

**Fig. 8.6** Spectrum of welding arc light: (a) Ar–$Co_2$; (b) TIG.

ing arc. The spectrum of welding arc light for the Ar–$CO_2$, TIG and plasma processes is shown in Figure 8.6. A semiconductor laser of wavelength 700 nm and narrow band filtering enable the laser light to be readily detected. Table 8.1 shows the application features of semiconductor lasers.

**Table 8.1** Application features of semiconductor lasers

| | |
|---|---|
| Oscillation | Semiconductor $p$–$n$ junction<br>Continuous oscillation<br>Pulse oscillation |
| Wavelength | 0.82–0.78 µm |
| Features | Easy modulation of laser<br>Small, lightweight |
| Applications | Photoswitch<br>Distance sensor<br>Form detection by laser scanning across joint |

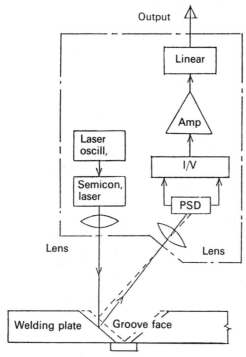

**Fig. 8.7** Scatter reflection type of laser sensor.

## 8.5 CONCLUSIONS

Conventional circuits for semiconductor laser distance sensors are shown in Figures 8.7 and 8.8. Figure 8.7 shows the scatter reflection type of laser

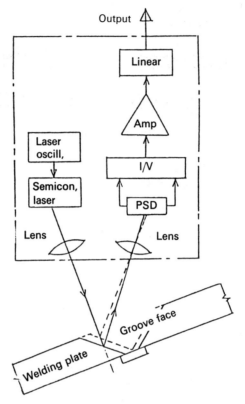

**Fig. 8.8** Reflection type of laser sensor.

sensor, in which the laser light is scattered by the joint face, and some is collected in the sensor, enabling satisfactory detection of distance. The reflection type of sensor (Figure 8.8) gives greater accuracy and is suitable for applications such as groove measurement, but it requires good reflecting surfaces. The characteristic output signal is linear, but variations in surface reflectivity cause large detection errors.

Safe operation of laser beams is controlled by established standards such as the US standard (FDA) or, in Japan, safety standard JIS C6802.

## REFERENCES

1. Inoue, K. *et al.* (1981) Image processing for on-line detection of welding process. Report 2: Improvement of image quality by incorporation of spectrum of arc. Japan Welding Society abstract no. 11.
2. Oomori and Ineka (1988) Foundation by laser sensors, in *Sensor Practical Dictionary*, Chapter 4.

# CHAPTER 9

# An automatic multilayer welding system with laser sensing

*Gouhei Iijima, Sumihiro Ueda,*

*Masaaki Hirayama, Kaoru Kawabe*

*and Osamu Takamiya*

The authors have developed the laser stripe optical sensor including not only shielding against the welding arc but also the control system to reduce the influence of undesirable light by both optical and software methods. Applying this sensing method to the fillet welding of cylindrical structure, an effective automatic multilayer welding system for joining thick-walled plates has been developed and put into practical application.

## 9.1  INTRODUCTION

In general, contact sensors are used for the automation of welding of thick-walled plate, for which the multilayer method has been mainly adopted. It is necessary to ensure the dimensional accuracy of the groove. To deal with irregularity of the groove, development of a sensor has been proposed in order to measure the groove and determine the correct position for welding. [1]–[4].

In this chapter, we describe a sensor for determining the welding position for each path of a multipass weld by a light section method using a laser and a TV camera. We report the results obtained from application of the laser sensor to automation of welding for a cylindrical structure.

*Sensors and Control Systems in Arc Welding.* Edited by Hirokazu Nomura.
Published in 1994 by Chapman & Hall, London. ISBN 0 412 47490 5

## 9.2   WORK PIECE

Figure 9.1 shows an example of a welding joint on the work piece and the appearance of the multilayer sequence. A support ring is butt-welded on the outside of the cylindrical object. Cold bending is excuted on the cylindrical object and gas cutting is performed on the support ring. The groove which is formed by the joint of the cylindrical object and the support ring is irregular.

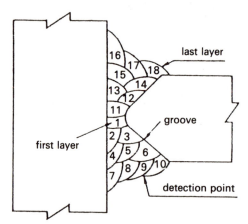

**Fig. 9.1** Example of welding joint and multilayer sequence.

## 9.3   DETECTION OF WELDING POSITION BY
## LIGHT SECTION METHOD

Figure 9.2 shows the principle of detection of the welding line by use of a light section method. A laser stripe is projected on to the groove and its image is captured at an oblique angle by a TV camera. The image from the TV camera is a bright line (light sectional line) adapted to the cross-section of the joint or weld. The image is decomposed into an appropriate number of pixels. Each pixel is classified in binary fashion as either white or black in accordance with a threshold luminance level. The threshold should be set so that only pixels that can be adapted to the light section method can provide clearness. Figure 9.2(b) shows the binary image. The welding position is presented as a concave feature. The middle point of the light sectional line is determined for each sample of the image taken from the binary image, and then the lines are subdivided. The slope of an individual straight line is obtained from the subdivided line drawing shown in Figure 9.2(c). The point at which the slope changes sign is considered as the centre point at which to aim the welding arc.

In multilayer welding, many paths exist and so multiple concave features appear in the image from the TV camera. The point on the profile at which

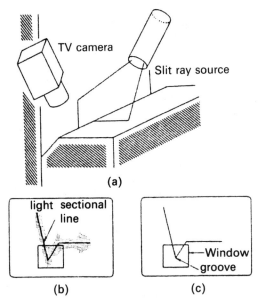

**Fig. 9.2** Detection of welding position: (a) principle; (b) binary image; (c) line drawing.

to aim is set before welding and identified by positioning a cross-line cursor on the TV monitor. If more than one concave feature appears in the window, the closest to the aim point is detected for processing.

## 9.1  MEASUREMENT OF THREE-DIMENSIONAL POSITION

Figure 9.3 shows principle of measurement of three-dimensional position. The laser stripe is irradiated on to the work piece to be measured. The TV camera captures the bright line (light sectional line) on the surface of the work piece. Transformation of the coordinate system of the camera position, $X_0$, and the coordinate system of the work piece, $X$, can be given by the following equation using a homogeneous coordinate system:

$$X_0 = C \cdot X \tag{9.1}$$

The coordinate system of the camera position is denoted by $X_0 = (hXx, hYc, hZc, h)^T$. The transform matrix is

$$C = \begin{pmatrix} C_{11} & C_{12} & C_{13} & C_{14} \\ C_{21} & C_{22} & C_{23} & C_{24} \\ 0 & 0 & 0 & 0 \\ C_{41} & C_{42} & C_{43} & C_{44} \end{pmatrix}$$

The coordinate system of the work piece is described by $X = (X, Y, Z, 1)^T$.

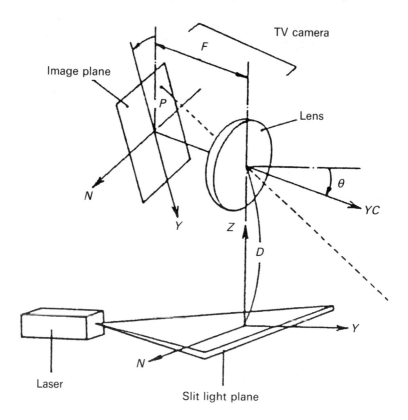

**Fig. 9.3** Principle of measurement for three-dimensional coordinates.

The individual elements of the transform matrix $C$ are called *camera parameters*. If the coordinate system $X$ of the work piece is determined so that the plane of the light stripe is denoted by $Z=0$ and the central point of the lens is set on the $z$ axis, element $Z$ of the light sectional line is always given by 0. Hence the values of the camera parameters necessary to measure three-dimensional position are given by

$$\begin{pmatrix} hXc \\ hYc \\ h \end{pmatrix} = \begin{pmatrix} C_{11} & C_{12} & C_{14} \\ C_{21} & C_{22} & C_{24} \\ C_{41} & C_{42} & C_{44} \end{pmatrix} \begin{pmatrix} X \\ Y \\ 1 \end{pmatrix} \tag{9.2}$$

The camera parameters can provide the distance $D$ between the light stripe and the lens, the slope and rotation of the stripe and the distance $F$ between the lens and the plane lit by the stripe:

$$\begin{pmatrix} C_{11} & C_{12} & C_{14} \\ C_{21} & C_{22} & C_{24} \\ C_{41} & C_{42} & C_{44} \end{pmatrix} = \begin{pmatrix} -\cos\alpha & \sin\alpha & 0 \\ \sin\alpha\cos\theta & -\cos\alpha\sin\theta & D\cos\theta \\ (1/F)\sin\alpha\cos\theta & (1/F)\cos\alpha\cos\theta & (D/F)\sin\theta \end{pmatrix} \tag{9.3}$$

## 9.5 PRACTICAL OPERATION

The laser sensor mounted in front of the welding torch determines the aim position during the welding operation. In practice, the greatest difficulty is caused by the effects from disturbing light such as the arc light produced during welding. A shield is installed as protection between the torch and the laser sensor in order to avoid spatter. In spite of the protection it is impossible to shield the sensor completely.

### 9.5.1 Automatic threshold adjustment

Figure 9.4 shows the method used to measure profile threshold. Under the following conditions, detection of threshold is not possible. First, the binary image may not be viewed as a light stripe if excessive numbers of ambiguous white pixels are present. Second, a concave feature may not be recognised in the line drawing if there is an inadequate white image. The prior threshold is utilized as a substitute if the above condition continues for a definite time.

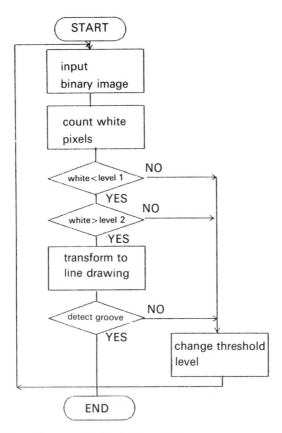

**Fig. 9.4** Flowchart of automatic threshold control.

The binary value is changed in order to keep the amount of white image at a suitable level if the threshold is absent for longer than the predetermined time.

### 9.5.2   Optical filter

Arc light is intense, and its spectrum of wavelengths is widely spread. Laser light is monochromatic. If a filter of appropriate wavelength is used, it can minimize the effect of disturbing light.

Figure 9.5 shows the relative radiant intensity of the laser and specification of the spectrum provided by the inserted filter for laser sensor. Both can eliminate or reduce the effect from disturbing light because they possess peaks around 780 nm wavelength.

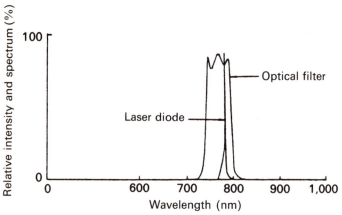

**Fig. 9.5** Spectrum characteristic of sensor.

Because the joint groove is machined in order to provide a more convenient condition for welding, the laser light reflects and halation appears on the TV camera, making it difficult to measure the aim point. An optical filter is therefore fitted inside the sensor to reduce disturbances generated by halation.

Figure 9.6 shows the input image before the above strategies are effected, and Figure 9.7 shows the binary image afterwards. The measurement of the aim point is represented by crossed lines.

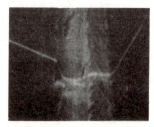

**Fig. 9.6** Image without optical filter.

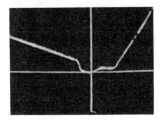

**Fig. 9.7** Example of welding position detection.

## 9.6  CONFIGURATION OF WELDING SYSTEM

Figure 9.8 shows the structure of the sensing head, and Table 9.1 presents the specification of the laser sensor. A mirror is equipped below the CCD camera to bend the light path so that the sensing head can be miniaturized.

Figure 9.9 shows the welding head, and Figure 9.10 indicates the configuration of the complete welding system. The microcomputer for profile control provides the following functions:

1. sampling of image signal;
2. measurement of aim position;
3. transformation of aim position into three-dimensional coordinates;
4. position control of welding torch;
5. communication with host controller.

We now discuss the application of the laser sensor to welding. The welding torch and sensor are mounted on a manipulator which is interfaced

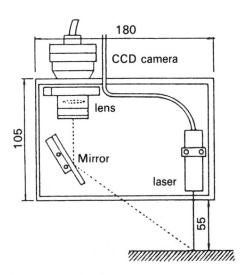

**Fig. 9.8** Configuration of sensor head.

**Table 9.1** Specification of laser sensor

| | | |
|---|---|---|
| Camera | Device | 2/3 inch CCD |
| | Number of pixels | $510 \times 492$ |
| | Scanning method | 2:1 interlace |
| | Synchronization | Internal and external |
| | Video output | Composite video out IVp-p 75 Ω |
| Laser | Device | Laser diode |
| | Wavelength | 780 nm |
| | Optical power | 1 mW |
| | Slit thickness | Under 1.0 mm |
| Other | Sampling time | 280 ms |
| | Resolution power | 0.3 mm |
| | Measuring range | Over 52.0 mm $x$ axis |
| | | Over 43.7 mm $z$ axis |
| | Dimensions | $180 \times 105 \times 45$ mm |
| | Weight | 700 g |

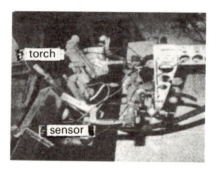

**Fig. 9.9** View of welding head.

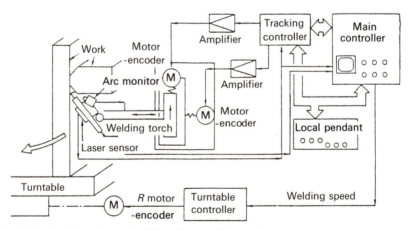

**Fig. 9.10** Configuration of welding line tracking unit.

to a work positioner having two axes with encoders and driven by servomotors. The work to be welded is attached to the positioner which can rotate about each axis at the required speed to provide work motion coordinated to that of the welding manipulator via commands from the host controller. When programmed and operating in automatic mode the welding torch is positioned at the programmed welding points. Any error between a taught position and the actual target joint position is measured by the sensor. The sensor measurement is then transformed into encoder values which are compared with the positioner's current encoder values. This provides a feedback signal to the servomotors so that the positioner moves the work to the correct position relative to the torch. Thus welding is automatically maintained on the required path.

## 9.7 CONCLUSION

Work is in progress to adapt this sensing technique to measurement of welding bead position, and to develop automatic failure detection using ultrasonics running along the welding bead.

## REFERENCES

1. Wada *et al.* (1986) *Mitsubishi Jukou Giho* **23** (2).
2. Inoue, K. (1980) *Quarterly Journal of Japan Welding Society*, **49** (9).
3. Inoue, K. (1981) *Quarterly Journal of Japan Welding Society*, **50** (11).
4. Masaki, I. (1981) *KHI Technical Review* No. 78, Aug.

# CHAPTER 10

# On-line visual sensor system for arc-welding robots

*Makoto Endo*

This paper describes the development of a visual sensor system for arc-welding robot, using a scanning, range-finding laser spot sensor. The author has developed a light-weight, compact optical sensor. It is incorporated in a system that can measure a welding groove on line without interference from arc light, and can capture the profile of the work piece to be welded. The system calculates the welding point and the three-dimensional vector to correct the welding robot. The values of the vector are fed back to the controller of the robot to realize automatic on-line seam tracking.

## 10.1  INTRODUCTION

Laser spot scanning or laser stripe projection are normally used in applying visual sensors to arc-welding robot. However, it is difficult to measure the work piece near the welding torch in the optically noisy conditions caused by strong arc light, particularly in high-current arc welding. In this sensor system, the following method is introduced to get over the difficulty. A laser beam, modulated at high frequency, is projected on to the work and detected by a highly responsive photodiode to convert it to an electrical signal which is passed through filters and detective circuits. In this way the sensor signal within the arc light is detected with an excellent S/N ratio. Thus we can obtain information regarding the position of the point or weld pass within a joint so that the welding robot can correct its path for seam tracking.

*Sensors and Control Systems in Arc Welding.* Edited by Hirokazu Nomura.
Published in 1994 by Chapman & Hall, London. ISBN 0 412 47490 5

## 10.2 PRINCIPLE AND CONSTRUCTION

### 10.2.1 Principle of measurement

The spectrum of arc light in welding is widely distributed from the ultraviolet to the infrared. The energy does not fall uniformly in all areas. If an optical filter is appropriate to the wavelength of the laser sensor, the S/N ratio is improved slightly, but not enough. However, time–frequency shows that the energy of arc light is considerably attenuated above 100 kHz. A high-frequency modulated laser beam is therefore adopted as the signal beam in this system. The signal is processed by detective circuits after conversion to an electrical signal. It has thus been possible to detect the laser light within the arc light with a high S/N ratio.

The method of measurement is as follows. A spot of modulated laser light is scanned on to the surface of the work piece. Then the vertical and horizontal positions are calculated by triangulation from the projection angle of the spot and the position of the spot on the image-formatted surface. The profile of the work piece is obtained by scanning the spot beam continuously across the surface, using an oscillating mirror (galvanometer).

Figure 10.1 shows the principle of measurement.

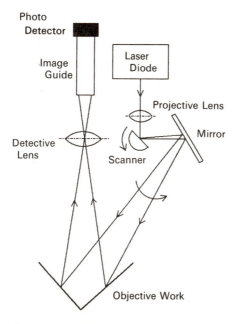

**Fig. 10.1** Principle of measurement.

### 10.2.2   Construction of sensor system

Figure 10.2 shows the construction of the sensor system.

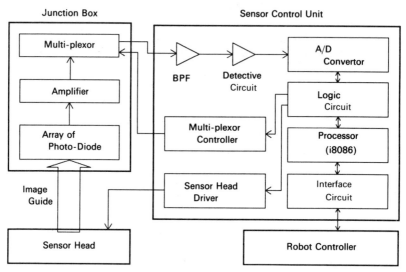

**Fig. 10.2** Construction of sensor system.

#### (a)  Sensor head

This both projects and detects the laser beam for measurement. The source of the beam is a laser diode, operating at a wavelength of 830 nm and a power of 30 mW. The beam from the diode is converged by a collimator lens and scanned by a galvanometer mirror. The image of the spot is formatted by the detector lens, and conducted to the junction box via the fibre-optic image guide.

#### (b)  Junction box

The signal beam conducted via the image guide is converted to an electrical signal by an array of 64 photodiodes. The output signal is amplified and multiplexed by the control signal from the sensor control unit.

#### (c)  Sensor control unit

This unit controls the laser diode, galvanometer mirror and multiplexer. The electrical signal from the junction box goes through a bandpass filter, detector circuits and an A/D convertor. The internal processor (i8086) calculates two-dimensional profile data after A/D conversion. The result of calculation is transferred to the robot controller through the serial interface circuit.

## 10.3 PERFORMANCE

The specifications and performance of the sensor are listed in Table 10.1. Figure 10.3 shows overview of the sensor head.

**Table 10.1** Specifications and performance of sensor

| | |
|---|---|
| *Specifications* | |
| Size of sensor head | Length 79.0 mm |
| | Width 69.0 mm |
| | Height 35.0 mm |
| | Weight 450 g |
| Field of view | 29.0 × 21.0 mm |
| Distance between head and work | 118 mm |
| Measurement position | 20 mm forward from welding torch position |
| *Performance* | |
| Accuracy of measurement | 0.4 mm |
| Function | Detection of start and end of edge |
| | On-line automatic seam tracking |
| Applicable work | Single V groove |
| | Fillet joint |
| | Lap joint |
| | Butt joint |
| Welding position | Free |
| Welding current | Maximum 1000 A |

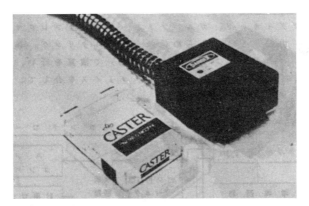

**Fig. 10.3** View of sensor head.

### 10.3.1 Special features

1. The sensor is able to measure in optically noisy conditions such as arc light.
2. As the sensor head is light and compact, it causes little restriction of robot movement.
3. The sensor head is specially protected from spatter to enhance its resistance to adverse operating environments.

4. Measurement is not affected by the profile of the material of the work piece, or the welding conditions.

### 10.3.2 Examples of measurement

Figure 10.4 shows measurement of a lap joint in 3 mm thick plate. Figure 10.5 shows measurement of a fillet joint. The dot indicates the result of the calculated welding point.

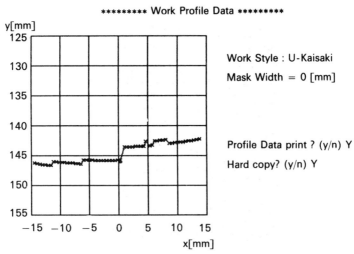

**Fig. 10.4** Measurement of lap joint.

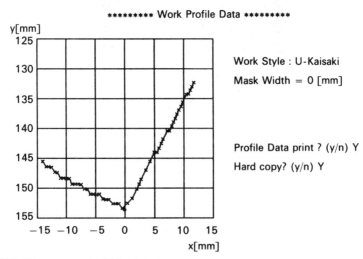

**Fig. 10.5** Measurement of fillet joint.

The computer in the robot controller calculates the three-dimensional vector required to correct the welding robot, so that it executes automatic seam tracking.

In this sensor system, the sensor system receives a measurement start command, detects and measures the work surface, and calculates the work profile and the three-dimensional vector. This process takes about 150–200 ms.

Figure 10.6 shows the welding robot, equipped with the sensor, executing on-line automatic seam tracking. Figure 10.7 shows a welded testpiece after seam tracking. In this case the robot was taught only two points of both edges of the work.

**Fig. 10.6** Robot executing seam tracking.

**Fig. 10.7** Testpiece after welding.

### 10.4 CONCLUSION

By applying this sensor system, the welding robot can be used without consideration of heat distortion during arc welding, and without accurate fixing and location of the work. Since the system can execute automatic seam tracking in high-current welding, it is applicable to most welding work from thin plates to large construction materials. Development of this system will be a key component of the realization of more intelligent robot systems.

## REFERENCES

Summaries of Fourth Annual Conference of Robotics Society of Japan. (No. 2602 Dec. 1986).

# CHAPTER 11

# Welding robot with visual seam tracking sensor

*Akira Hirai, Nobuo Shibata, Toshio Akatsu and Kyoichi Kawasaki*

A welding robot has been developed that corrects the taught path automatically in accordance with seam line deviation detected by a visual sensor. The optical component of the sensor head is set at a distance from the torch arc point to avoid contamination of the optical component by fumes or spattering, and interference between sensor head and workpieces or jigs. An image processor processes the image data quickly and accurately by compressing it to a local value point address of intensity difference. This system is suitable for $CO_2$ or MAG arc welding on thin (0.8–4.5 mm) sheet steel, and it simplifies the production of high-quality welds.

## 11.1  INTRODUCTION

Current research seeks to provide a robot arm and control with sensor that functions in an environment in which the desired path is not predictable, nor repeatable from one cycle to another.

To achieve this goal many types of welding sensor have been developed; for example, the arc-current-controlled (ACC) sensor and the magnetic proximity sensor [1]. But these sensors restrict the range of applicable joint types and welding methods.

A visual sensor, however, can detect thin plate joints (below 2 mm) and it does not require weaving. Many types of visual sensor have been developed [2]–[4].

*Sensors and Control Systems in Arc Welding.* Edited by Hirokazu Nomura. Published in 1994 by Chapman & Hall, London. ISBN 0 412 47490 5

This chapter describes a visual sensor system designed not to restrict welding flexibility. The system uses triangulation range finding to detect seam position. To achieve availability and reliability in manufacturing, we have improved the optical component arrangement, image processing algorithms, and robot control method using sensor data [5].

## 11.2   SYSTEM OVERVIEW

The prototype system (see Figure 11.1) consists of an industrial welding robot, welding power supply, robot controller, sensor head with welding torch, and an image processor. The robot is a Hitachi M6060 welding robot with six degrees of freedom of motion. The welding power supply is Hitachi SE-350, and the welding method is $CO_2$ arc welding. The welding torch is fixed to the robot's wrist. The sensor head is arranged around the welding torch. The distance between the optical window of the sensor head and the welding arc point is large enough to eliminate the influence of fumes and spattering caused by welding. The reflected image of the sensor's laser stripe from the work surface can be obtained clearly for a long time without cleaning the window. The sensor does not obstruct the torch position at any position or attitude, even if positioning or fixing tools are present.

The robot controller is a multiprocessor system. A servo control processor controls the joint angle of each arm, and the welding condition. A sequence control processor is responsible for managing the system operation. A communication control processor is used to communicate with the image processor, to transfer detected coordinate data and to correct the tracking path using these data. Each processor transfers data via a bus line.

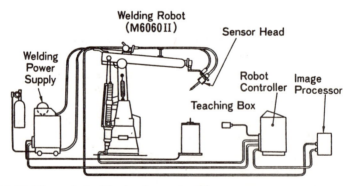

**Fig. 11.1** Welding robot with visual seam tracking sensor.

## 11.3   SENSOR HEAD CONFIGURATION

The sensor head assembly is illustrated in Figure 11.2. The sensor head comprises a light stripe emitting block and an image-detecting block. The

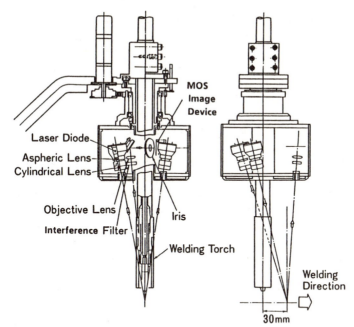

**Fig. 11.2** Construction of sensor head.

light stripe emitting block includes a pair of Ga–Al–As infrared laser diodes combined with an aspheric lens and a cylindrical lens. The laser diode is packaged with a perche cooling module and a thermistor. The temperature of the laser diode is maintained within a constant temperature range to minimize wavelength fluctuation of the laser diode. Light stripes from a pair of laser diodes and their optics are arranged to compose the same sheet plane of light. The laser light beam is at an angle to the welding seam line to avoid coincidence of the single reflected light image from the workpiece with double reflected images caused by mutual illumination of one surface by reflection from another.

The image-detecting block has a Hitachi MOS image device (HE97211) combined with an object lens, interference filter and optical iris. The image is reduced to one third the size of original.

The MOS sensor detects the image produced by the light stripe projected on the surface of the workpiece through the window, the iris and the interference filter. By observing the stripe image through the interference filter, the disturbing influence of the welding arc is removed. The video signal from the image device is digitized and processed by an image processor.

The sensor detecting region is about 30 mm in front of the arc point of the welding torch. The observation direction from the torch can be controlled using the geared d.c. motor.

There are three cylindrically shaped openings with windows in the frame body. Purge air blown in the openings prevents spatters and fumes from adhering to the surface of the windows.

## 11.4   IMAGE PROCESSING

The captured images of the stripes are fed to the image processor. The image processing flow is shown in Figure 11.3. The processing comprises hardware-based preprocessing, and software-based post processing. Image data from the sensor is compressed into quasi-binary address data in the pre-processing. An analogue-to-digital converter samples and quantifies the video signal from the MOS image device, and sends the signal to the pre-processing hardware. This calculates the average difference of five adjacent pixels and writes the raster address where the average difference is a local maximum or minimum.

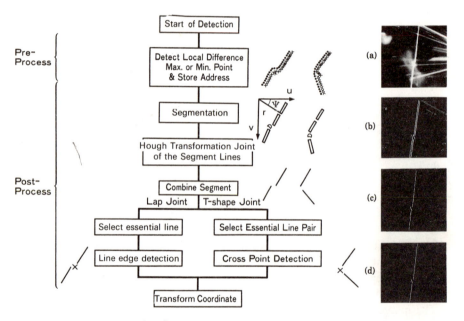

**Fig. 11.3** Image-processing flow.

By using this circuit, edge pairs of the line along the raster directions are obtained, and the moderate change in brightness caused by the strong illumination of the welding arc is eliminated (see Fig. 3(b)).

At the next stage, a software algorithm separates the image into segments according to continuity judgement of the quasi-binary address data. Then Hough transformation and line connection of adjacent line pairs are executed. As a result of this process, the essential lines that satisfy the properties of joint models are successfully selected.

Finally, the edge of the lower surface stripe image (for a lap joint) or the crossing point (for a T-shape joint) is computed. This point becomes the seam position on the image plane.

3. The robot controller only modifies the taught line according to the seam line position detected by the vision sensor, so that a combination of the sensing and non-sensing regions is easily achieved.

## REFERENCES

1. Araya, T., Tsuji, M. and Aro, T. (1985) Advanced technology for arc welding robots. *Hitachi Review*, **34**(1), 27–32.
2. Bamba, T., Muruyama, H., Codaira, N. and Tsuda, E. (1984) A visual seam tracking system for arc-welding robot, in *Proc. 14th Int. Symp. on Industrial Robots*, pp. 365–74.
3. Corby, N.R. Jr (1984) Machine vision algorithms for vision guided robotic welding, in *Proc. Fourth Int. Conf. on Robot Vision and Sensor Controls*, pp. 137–47.
4. Nakata, S. and Jie, H. (1989) Construction of visual sensing system for in-process control of arc welding process and application in automatic weld line tracking. *Journal of the Japan Welding Society*, **4**(4), 45–50.
5. Hirai, A., Shibata, N., Akatsu, T. and Tomita, M. (1987) Welding robot with visual seam tracking sensor, in *Proc. USA–Japan Flexible Automation Symposium*, pp. 1055–1060.

# CHAPTER 12

# An arc-welding robot with a compact visual sensor

*Takao Bamba*

This chapter describes an arc-welding robot system with a new type of visual sensor designed to track welding seams with greater precision and under more diverse conditions than previous sensors. The sensor, consisting of a laser diode and a photodetector, is mounted directly on the welding torch and employs triangulation to detect the welding seam. The system features on-line tracking, simplified teaching, and flexible compensation for workpiece positioning error and thermal distortion.

## 12.1   INTRODUCTION

Because of the recent progress of industrial robots, many robotic welding systems have been introduced to the factory shop floor. A considerable amount of research has been carried out on robotic welding systems with advanced sensors, because they need adaptive control technology for precise and smooth tracking motion along the three-dimensional seams of work-pieces [1]–[5].

This chapter describes our compact visual sensor and its control system suitable for an articulated arc-welding robot with six degrees of freedom, which we developed in 1985 [8], [9].

## 12.2   OPERATING PRINCIPLE AND CONSTRUCTION

We have designed a visual sensor specifically for articulated arc-welding robots. In such a mechanism it is necessary to measure the relative

*Sensors and Control Systems in Arc Welding.* Edited by Hirokazu Nomura.
Published in 1994 by Chapman & Hall, London. ISBN 0 412 47490 5

~-dimensional distance between a seam and the torch from any welding
po. n along various workpieces. It is therefore preferable to use a
non- ontact sensor to avoid interference with them. We adopted an optical
method which is well known as a light section method based on triangulation [6]. It is a technique for obtaining range information by projecting a
stripe light pattern on an object, and detecting the stripe image on the object
from different angles.

In our sensor, the sensing method uses the light section method with a
distance-detection unit integrated compactly with the welding torch, and
oscillated around the torch as shown in Figure 12.1. To detect a seam
position, the visual sensor samples distance detection values as required in
synchronization with this scanning period, and analyses the distance image
obtained using a microprocessor. For example, the weld groove shapes
illustrated in Figure 12.2 have been detected as the patterns indicated at the
bottom of the figures. If the break point of each pattern is recognized, the
seam position can be defined.

Figure 12.3 shows the construction of the distance-detection unit. It has
a projector and a receiver which are arranged so that their beam axes cross
at a predetermined angle $\theta$. Beams output from the projector are focused by

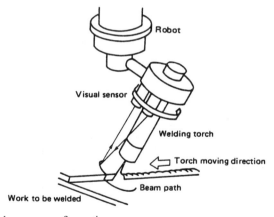

**Fig. 12.1** Visual sensor configuration.

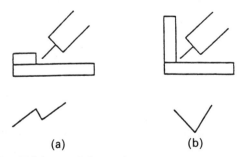

(a)          (b)

**Fig. 12.2** Typical weld joints and detected patterns.

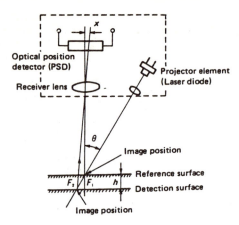

**Fig. 12.3** Configuration of distance-detection unit.

the optical system into a minute light image (0.5 mm in diameter) on the surface of the work to be welded. By detecting this image position $x$ with a unidimensional optical position detector, the distance displacement $h$ from a beam axis crossing point $\Gamma 1$ can be obtained. Since a small light source with high brightness is required to ensure on-line position detection, the projector uses a high-output laser diode (30 mW light output, 830 nm wavelength). We use a prism in the projector to bend the laser beams so as to make the distance measuring unit as compact as possible. The sensing element comprising the receiver is a one-dimensional semiconductor position-sensitive detector (PSD). This semiconductor element has an even resistance layer on its surface and is capable of detecting an image position in terms of an analogue quantity from the optical current which is generated at the image position on the receiving surface. Therefore, unlike a solid-state image element, this semiconductor element does not have a dead zone between picture elements, and allows position detection with high accuracy. The distance-measuring unit can be driven round the welding torch by a stepping motor for scanning. The mechanism is effective for controlling the centre of the scanning motion during seam tracking.

Figure 12.4 shows the hardware configuration of the sensor. A 16-bit microprocessor used for visual sensor data processing executes the distance-detection value sampling, shape data analysis, seam position determination, calculation of robot motion compensation, and various status settings. This processor also controls the stepping motor, which acts as a visual sensor oscillation drive actuator, to perform seam tracking control round the welding torch axis.

The projector and receiver are interfaced with the microprocessor by a laser power driver and PSD signal-processing circuit. To eliminate noise due to disturbance light, the laser diode output is pulse modulated and de-modulated by the PSD signal-processing circuit.

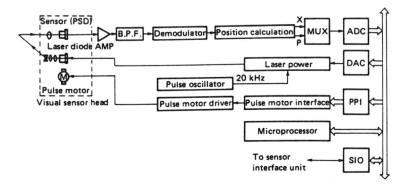

**Fig. 12.4** Configuration of visual sensor hardware: BPF, bandpass filter; ADC, A/D converter; DAC, D/A converter; PPI, programmable peripheral interface; SIO, serial input/output controller; MUX, multiplexer; PSD, position-sensitive detectors.

Figure 12.5 shows the welding torch equipped with the visual sensor. The projector and receiver are two lens tubes protruding from the disc-shaped torch holder (white area) located at the top of the welding torch. Since the projector and receiver lens tube ends are designed to be approximately 150 mm away from the workpiece surface to be welded and also the detection unit is compact, torch accessibility to joints is not greatly restricted. Table 12.1 shows the specification of the visual sensor, and Table 12.2 shows the applicable joints.

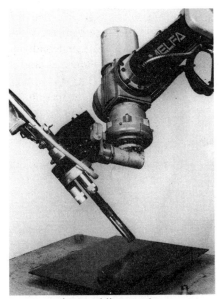

**Fig. 12.5** Visual sensor mounted on welding torch.

**Table 12.1** Visual sensor specification

| | |
|---|---|
| Stand-off | $150.0 \pm 0.1$ mm |
| Field of view from stand-off | $\pm 7.0$ mm |
| Precision | $\pm 0.2$ mm |
| Laser output | 30 mW |
| Diameter of laser spot | 0.2–0.5 mm |
| Projection angle | $13.44° \pm 0.5°$ |
| Sensing point from torch centre | 28.0 mm |
| Scanning angle | 20° |
| Scanning area | 200° |
| Scanning rate | 4 Hz |
| Scanning resolution/pulse | 0.12 mm |
| Required air | Dry air 2–3 kg/cm2 |
| Weight | 2 kg |

## 12.3  PROTECTION FROM ARC-WELDING ENVIRONMENTS

For correct operation during arc welding, it is necessary to prevent spatter generated during the welding process from attaching to the optical lens, interruption of the light path due to fumes, and saturation of the receiver elements due to strong arc light.

In this visual sensor, since spatter is generated at the welding area of the torch wire tip and spreads radially, and the sensor head has been installed above the welding torch so that this sensor head is in the shade of the torch itself, the probability of spatter reaching the optical system is small. Nevertheless, since some spatter is reflected from the work surface and flies to the sensor, the optical system is provided with an interchangeable filter. By changing this filter periodically, sensor protection is achieved. The sensor is protected from fume and heat by adopting a design in which air guided within the lens tubes flows across the end of each lens tube.

## 12.4  ROBOT CONTROL

The robot control system has two main functions, as follows.

### 12.4.1  Finding start point

Before welding, the robot is positioned at a preprogrammed start point where target sensing and data modification by the measured results are performed. Because of the offset of the sensing point and the actual welding point, the torch is first moved to the offset point where the sensing point coincides with the preprogrammed point and then to the modified start point.

**Table 12.2** Applicable joints

| Name | Condition | | Description |
|------|-----------|---|-------------|
| 1. Lap | Shape | | Gap: Sheet thickness or less |
| | | | Edge radius: 1/4 of sheet thickness or less |
| | | | Edge angle: 90° |
| | | | Forward angle: 10° or less |
| | | | Backward angle: 10° or less |
| | | | Gap: 1/3 of sheet thickness or less |
| | | | Edge radius: 1/4 of sheet thickness or less |
| | | | Edge angle: 90° |
| | | | Forward angle: 10° or less |
| | | | Backward angle: 10° or less |
| | Material | | SPHC, SPCC, SS materials |
| | | | Material should not greatly change the light reflection ratio of plating, surface treatment, etc. |
| | Number of detections | | There should be only one of the same type joint within detection range. |
| 2. Fillett | Shape | | Gap: 3 mm or less |
| | | | Bottom angle: 90° |
| | | | Forward angle: 10° or less |
| | | | Backward angle: 10° or less |

**Table 12.2** *contd*

| | | |
|---|---|---|
| | Material | SPHC, SPCC, SS materials<br>Material should not greatly change the light reflection ratio of plating, surface treatment, etc. |
| | Number of detections | There should be only one of the same type joint within detection range. |
| 3. Corner | Shape | <br>Gap: 1 mm or less<br>Edge radius: 2 mm or less<br>Edge angle: 90°<br>Forward angle: 10° or less<br>Backward angle: 10° or less |
| | Material | SPHC, SPCC, SS materials<br>Material should not greatly change the light reflection ratio of plating, surface treatment etc. |
| | Number of detections | There should be only one of the same type joint within detection range. |
| 4. Butt disable | Shape | |
| 5. V disable | Shape | |

### 12.4.2 Seam tracking

This is a real-time tracking function which modifies the welding path autonomously based on the visual sensor ahead of the welding torch. Since the path-correction function uses a kind of preview control, the robot can continue to weld even where the sensor loses its way because of tack welds. In addition, as the sensor-positioning system is independent from the torch-positioning system, the sensor can be driven to the next direction when the seam path has a corner.

## 12.5 APPLICATION

We applied this sensor-based arc-welding robot to the seam weld shown in Figure 12.6. It is a component of a car body. The shape of the weld joint is not simple. There is also positional uncertainty. In general, our visual sensor is suitable for application to such a complex joint. Figure 12.7 shows a tracking trajectory by the visual feedback control. In the figure, the robot initially finds its start point, and then tracks precisely after start point modification. The width of the weld joint is approximately 1.0 mm, and the thickness of the sheet metal is 0.8 mm. The tracking error is within $\pm 0.4$ mm.

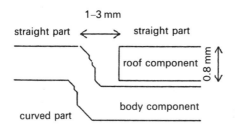

**Fig. 12.6** Cross-section of joint.

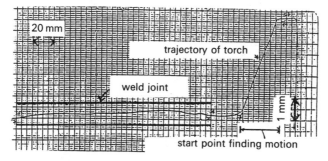

**Fig. 12.7** Tracking trajectory.

## 12.6  CONCLUSION

This chapter has described the on-line visual sensor and the control functions of the sensor-based robot. The system has the following features.

1. It is an on-line sensor which allows detection of the seam during welding.
2. The sensor head is compact to increase the welding torch's access to the workpiece.
3. High-detection resolution allows application to sheet welding.
4. It is applicable to various types of joint.
5. It is protected from fume, spatter, etc.
6. Short visual data processing permits use for high-speed welding.

## REFERENCES

1. Nicolo, V. *et al.* (1980) Industrial robots with sensory feedback application to continuous arc welding, in *Proceedings of Tenth ISIR*, Milan, pp. 15–22.
2. Cook, G.E. *et al.* (1981) Micro-computer control of an adaptive positioning system for robotic arc-welding, in *Proceedings of IEEE IECI*, pp. 252–68.
3. Morgan, C.G. *et al.* (1983) Visual guidance techniques of robot arc welding, in *Proceedings of Third Int. Conf. RoViSec*, Cambridge, pp. 615–24.
4. Kremers, J. *et al.* (1983) Development of a machine-vision-based robotic arc-welding system, in *Proceedings of 13th ISIR*, Chicago, pp. 14.19–14.33.
5. Baudot, W. *et al.* (1983) Visually guided arc-welding robot with self-training features, in *Proceedings of 13th ISIR*, Chicago, pp. 10.13–10.30.
6. Bamba, T. *et al.* (1981) A visual sensor for arc-welding robots, in *Proceedings of 11th ISIR*, Tokyo, pp. 151–8.
7. Kodaira, N. *et al.* (1983) Microcomputer control of an arc-welding robot with visual sensor, in *Proceedings in IEEE IECON*, Chicago, pp. 93–6.
8. Bamba, T. *et al.* (1984) A visual seam tracking system for arc-welding robots, in *Proceedings of 14th ISIR*, pp. 365–74.
9. Bamba, T. *et al.* (1985) An arc-welding robot with an on-line visual sensor. *Mitsubishi Denki Giho*, **59**(10), 705–708 (in Japanese).

# CHAPTER 13

# A robot with visual sensor for three-dimensional multilayer welding

*Teppei Asuka, Jyunzou Komatu, Kouichi Satou*
*and Michio Kitamura*

This chapter describes an accurate and widely applicable tracking control system which guides a welding torch along a joint line. It includes a robot controller, a vision processor and a weld data processor. The system detects the outline of a weld joint using He–Ne laser and a TV camera in the peak value vision technique, and a Vision Processor to calculate the position of the next pass. By using the next pass position and welding data, the welding conditions and the articulation angle of robot are decided and a weld control program is produced. This robot has been applied to welding of a large-scale steel structure and achieved improved welding quality and labour efficiency, and also showed its practicality.

## 13.1 INTRODUCTION

Nowadays, almost all robots used in automatic welding work have a sensor mechanism, and various sensors have been developed for use in the welding of thin plates. For welding of multilayers in thick plates, it is necessary to automate the correct positioning of a welding torch. This has led to the development of sensor techniques to check the multilayer surface conditions during welding, and a robot for three-dimensional multilayer welding with visual sensors.

*Sensors and Control Systems in Arc Welding.* Edited by Hirokazu Nomura.
Published in 1994 by Chapman & Hall, London. ISBN 0 412 47490 5

## 13.2   CONSTRUCTION OF EQUIPMENT

Figure 13.1 shows the construction of the equipment, which consists of a visual sensor to detect the welding conditions of multilayers in a weld joint, a processor to instruct both welding work and the positioning of a welding torch based on the bead surfaces indicated by the visual system, a robot mechanism controlled in accordance with the information from a weld data processor, and a welding machine.

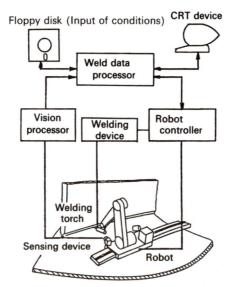

**Fig. 13.1** Construction of equipment.

### 13.2.1   Development of vision processor

#### (a)   Sensor

The outline of a weld joint obtained by projection of a light stripe from a He–Ne laser is photographically detected with a small TV camera. The laser is transmitted through fibre-optic cables and a filter is provided for the camera to avoid interference from the outside and to protect the visual system from any disturbance.

#### (b)   Vision processor

A horizontal histogram is produced by the vision system (Figure 13.2) which uses the peak value of brightness detected by the sensor as an outline. In this way, the error caused by the outline width of the light stripe is minimized. Abnormal points (due to interference from the outside) are neglected by adjusting the deviation of the outline (Figure 3.3) [1].

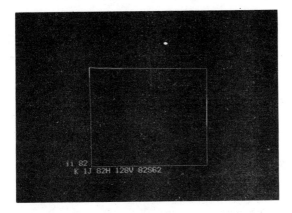

**Fig. 13.2** Two-value vision.

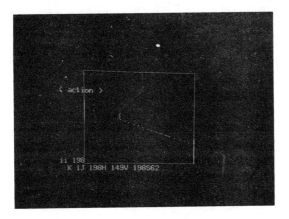

**Fig. 13.3** Peak-value vision.

## 13.2.2 Weld data processor

### (a) Control of positioning of welding torch

The concavities (A and B in Figure 13.4) on the beads are detected by comparing the shape of the current weld groove with that of the previous run after confirming the former by outline image. The upper concavity (A) is then decided using the sequence of multilayers, and the next pass position is decided. Figure 13.4 indicates the method of positioning of a welding torch.

### (b) Weld data processor

The weld data controller is closely related to the knowledge of multilayer welding. This controller will calculate the weld position for a robot from the aiming position for the welding torch determined from the sensed image and will produce welding requirements using the welding data input by a worker

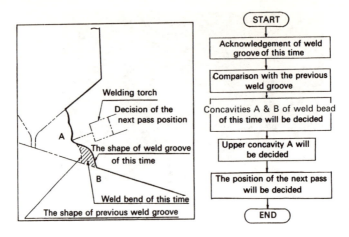

**Fig. 13.4** Control method for welding torch position.

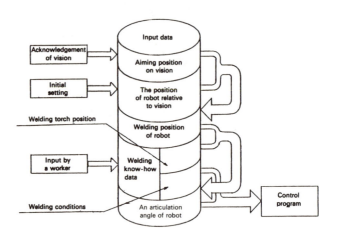

**Fig. 13.5** Construction of weld data control.

into a welding knowledge base. By using these welding data, welding torch position, welding conditions and the articulation angle of the robot will be decided and the welding robot control program will be produced. Figure 13.5 shows the construction of weld data control [2].

## 13.3   RESULTS

Figure 13.6 shows a macro-section of a weld. A satisfactory mutilayer weld was obtained in a double J-type joint of 70 mm thick plates.

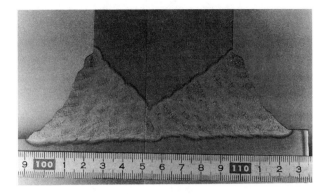

**Fig. 13.6** Typical weld.

## 13.4 CONCLUSION

By establishing the algorithm to decide the next pass in multilayer welding from the outline obtained from a laser stripe sensor, a robot has been developed for three-dimensional multilayer welding.

## REFERENCES

1. Hata, S., Ariga, M., Suzuki, K., Tsuji, M., Kawasaki, K. and Shimura, Y. (1984) 'Small image Processor HV/R-1 for robot vision.' *Hitachi Hyouron*, **66**(10), 735–40 (in Japanese).
2. Mori, S., Hata, S., Matsuzaki, K. and Takeda, K. (1986) 'FA-BASIC, a universal language for controlling factory automation cells,' *Hitachi Review*, **35**, 37–42.

# CHAPTER 14

# A visual arc sensor system

*Shinji Okumura and Seigo Nishikawa*

For greater improvement of reliability and quality in robotic welding, a visual arc sensor system has been developed that permits sensing in real time through arc-image processing. It uses binary image processing, and has made possible higher welding speeds (2500 mm/min), on thinner plate (1.0 mm), and detection for gap control.

## 14.1 INTRODUCTION

Visual sensors are essential for arc welding robots to be intelligent, but there are problems. Because the arc has a high intensity and a spectrum of wavelengths covering essentially the whole range, various approaches have been used. To solve these problems, in one, a molten pool vision sensor uses a high speed-shutter or a special filter to remove arc light from the image. An alternative visual sensor uses a laser which has a higher intensity than the arc and a different wavelength. But these methods are more complex and more expensive. A visual arc sensor (ARC-EYE) has been developed to enable sensing in real time through arc image processing, observing the arc itself without adding any special device. ARC-EYE's peripheral equipment is simpler and permits higher-speed sensing.

## 14.2 SYSTEM CONSTRUCTION

The block diagram of YASNAC-ERC (Yoskawa's Motoman robot controller) and ARC-EYE is shown in Figure 14.1. The vision control block of ARC-EYE is integrated with the robot control block, simplifying the construction and resulting in lower cost and higher speed.

*Sensors and Control Systems in Arc Welding*. Edited by Hirokazu Nomura.
Published in 1994 by Chapman & Hall, London. ISBN 0 412 47490 5

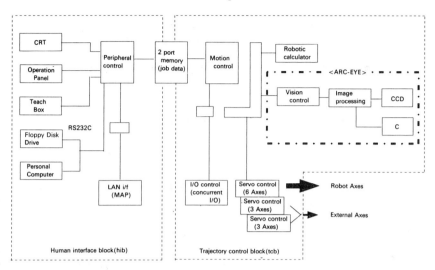

**Fig. 14.1** Block diagram of YASNAC-ERC and ARC-EYE.

## 14.3 PRINCIPLE

When an arc is struck on a flat plate, it forms an egg-shaped arc pattern as shown in Figure 14.2(a). However, when an arc is struck on a joint, it is interrupted by the upper plate and a pattern is formed as in Figure 14.2(b). There is a distinctive feature of a protrusion in the welding direction, called the arc top.

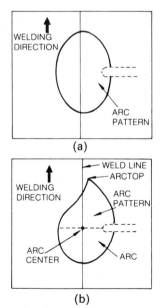

**Fig. 14.2** Arc pattern seen by camera: (a) flat plate; (b) joint.

The position of the arc top is always offset a certain amount from the weld line (root of the joint). When the workpiece has a deviation to the right or left of its nominal position, the position of the arc top will show a corresponding change. Also, the position of the arc centre will change in height.

### 14.3.1　Detecting deviations in the right–left direction

When there is no deviation the arc top is at the pre-taught weld line and no left/right corrections are performed (Figure 14.3(b). However, when the weld line deviates to the left of the pre-taught welding path as in Figure 14.3(a), the arc top is at the actual weld line and to the left of the pre-taught path. A correction operation is therefore performed in the left direction. In the same manner, when there is a deviation to the right, the position of the arc top is as in Figure 14.3(c) and a correction operation is performed in the right direction.

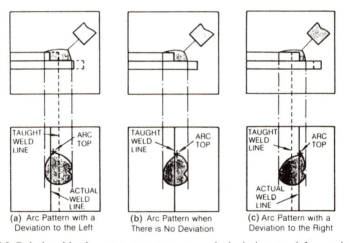

(a) Arc Pattern with a Deviation to the Left　　(b) Arc Pattern when There is No Deviation　　(c) Arc Pattern with a Deviation to the Right

**Fig. 14.3** Relationship between arc pattern and deviation to left or right: (a) deviation to left; (b) no deviation; (c) deviation to right.

### 14.3.2　Detecting deviations in the up–down direction

When there is no deviation in up or down, the arc centre is at the pre-taught weld line and no corrections are made (Figure 14.4(b)). However, when the actual weld line deviates in the down direction as in Figure 14.4(a), the arc centre is to the left of the pre-taught weld line and a correction operation is performed in the down direction. In the same manner, when there is a deviation in the up direction, the arc centre is as in Figure 14.4(c) and a correction operation is performed in the up direction.

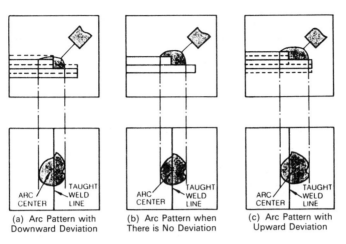

(a) Arc Pattern with Downward Deviation

(b) Arc Pattern when There is No Deviation

(c) Arc Pattern with Upward Deviation

**Fig. 14.4** Relationship between arc pattern and deviation up or down: (a) downward deviation; (b) no deviation; (c) upward deviation.

### 14.3.3 Gap control function

When there is no gap, an arc pattern is formed as in Figure 14.5(a). However, when there is a gap at the root of the joint, an arc pattern is formed as in Figure 14.5(b). The arc is not struck on the upper plate. As a result, the gap can be detected through the arc width $(l_1 - l_2)$ in real time by comparing the width with no gap and the width with a gap.

After the detection of gap and arc width, a correction operation is performed in the up direction to ensure proper fusion of the top plate and the welding speed is reduced according to predetermined data to ensure proper filling of the joint.

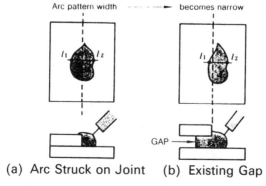

Arc pattern width ----- ► becomes narrow

(a) Arc Struck on Joint   (b) Existing Gap

**Fig. 14.5** Relationship between arc pattern on joint and arc width: (a) arc struck on joint; (b) gap present.

## 14.4   TRACKING SPECIFICATION

The tracking specification is shown in Table 14.1.

**Table 14.1** Seam tracking function

| | |
|---|---|
| Welding condition | |
| Gas | $Ar + CO_2$, $Ar + O_2$, pulse MAG |
| Wire | Solid wire (0.8–1.4 mm dia), core wire (1.2 mm dia) |
| Speed | 2500 mm/min (max) |
| Current | 90–500 A |
| Joint configuration | Lap joint                    Fillet joint |
| Workpiece | |
| Material | Mild steel, stainless steel |
| Fit up (gap) | Maximum gap half thickness (thickness 2.3–4.5 mm) Maximum gap 3 mm (thickness more than 4.5 mm) Note: welding quality (bead) is slightly low |
| Tracking angle | TAUGHT PATH    $\theta = 5.0°$ Weld speed 2500 mm/min Correction 0.15 mm |
| Features | Not affected by welding heat, spatter or fumes (air sweeper) Protective glass changed every six days |

## 14.5   CONCLUSION

Binary-image-processed arc patterns ensure accurate and positive detection. The arc pattern has another distinctive feature at the corner of the joint. In the near future we shall apply this system to more complex work and other joints.

# CHAPTER 15

# Fuzzy control of $CO_2$ short-arc welding

*Kenji Ohshima, Satoshi Yamane, Haruhiko Iida,*
*Shujiang Xiang, Yasuchika Mori and*
*Takefumi Kubota*

This chapter deals with some problems concerning the sensing of weld pool phenomena and the effects of power source characteristics on the stability of the arc in a $CO_2$ short-arc welding system. A power source characteristic is proposed for improving the stability and self-regulation of the arc and molten metal transfer. We have developed a non-linear power source consisting of transistors and function generators and have used it to perform experiments. A new approach is developed to obtain optimum welding parameters by fuzzy inference.

## 15.1   INTRODUCTION

It has been difficult to observe the weld pool with a TV camera, since the arc light is too strong. The authors tried to watch the weld pool without arc at the period synchronous with short circuiting of $CO_2$ short-arc welding.

In order to construct an artificial intelligent robot, it is important that the effect of spatter is removed. For this purpose, the control method of the short circuit and of the arc current and the effect of the power source characteristic on the stability of metal transfer are discussed. The validity of the theoretical results has been verified by experimental investigations. A power source characteristic is proposed for improving the stability and self-regulation of the arc and molten metal transfer in short-arc welding. The optimum welding parameters are obtained by fuzzy inference.

*Sensors and Control Systems in Arc Welding.* Edited by Hirokazu Nomura.
Published in 1994 by Chapman & Hall, London. ISBN 0 412 47490 5

## 15.2  DIGITAL CONTROL SYSTEM

The system is designed as shown in Figure 15.1. A CCD camera is used to watch the weld pool. Figure 15.2 shows a chart of the timing by which the observation and control of the weld pool are achieved. The arc voltage $V$ becomes low in the short-circuit period. To obtain the weld pool image in the short-circuiting period without the arc light, it is necessary for the shutter of the CCD camera to open when the arc voltage goes low. The shutter is opened by the external trigger pulse. (EXT.TRIG.) which is produced from the torch voltage $V$ by a timing pulse generator. The CCD camera requires a charge time (shutter time) of 1 ms to capture an image and transfers the video signal to a frame pose, for 1/60 s after the vertical synchronizing signal VD which appears after the external trigger. In the following 1/60 s, a computer processes the image and determines the power source characteristic (welding current $i$ and torch voltage $V$) so as to control

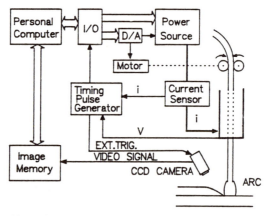

**Fig. 15.1**  The weld pool control system.

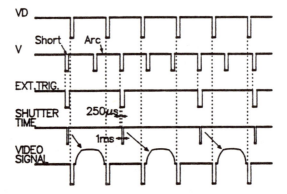

**Fig. 15.2**  Timing chart for observing the weld pool.

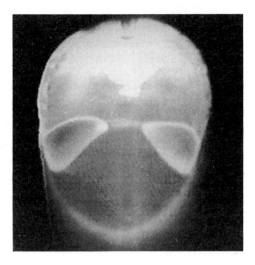

**Fig. 15.3** The weld pool image.

the weld pool shape. A typical image of the weld pool taken with the CCD camera is shown in Figure 15.3.

## 15.3 PHASE PLANE ANALYSIS

A new power source characteristic is proposed for improving the stability of metal transfer and for controlling the weld pool shape. The short arc is achieved by converting the power source characteristic between the short circuit and the arc characteristic. There are three kinds of power source characteristic: the power source characteristic 1 (PSC1) of the short circuit, the power source characteristic 3 (PSC3) of the arc, and the power source characteristic 2 (PSC2) when the short-circuit characteristic is shifted to a lower voltage. Phase plane analysis is used for the present investigation. The current $i$ and voltage $V$ constitute the coordinates of a representative point in the phase plane. Our purpose is to improve the stability of metal transfer and to control the short-circuit current and arc current independently. For this purpose, it is useful to investigate the trajectory of the solution in the transient states.

The experiments have been performed by using the new power source characteristic. The wire was mild steel of 1.2 mm diameter and the sealed gas was 100% $CO_2$. The trajectory of the $CO_2$ short-arc is plotted in the $i$–$V$ phase plane of Figure 15.4. The waveforms of the current $i$ and the arc voltage $V$ are shown in Figure 15.5. The investigation is begun from the steady state where the representative point lies at point 1 in Figure 15.4. Let the power source characteristic be PSC1. When the droplet at the top of the wire comes in contact with the base metal, the circuit resistance decreases

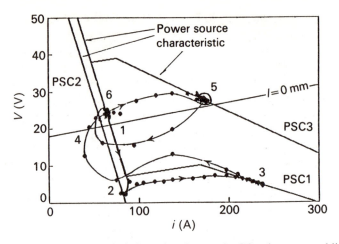

**Fig. 15.4** Power source characteristic and trajectory in CO$_2$ short-arc welding.

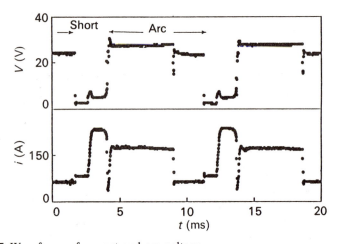

**Fig. 15.5** Waveforms of current and arc voltage.

and the welding torch voltage also decreases. The power source characteristic is converted to PSC2 from PSC1 so as to bring the droplet into sound contact with the base metal. The power source characteristic is then converted to PSC1 from PSC2 again, and the trajectory of the representative point $(i, V)$ moves to point 3 from point 2. The droplet transfers to the weld pool from the wire, and the current surges to the short-circuit value of PSC1 at the interface between the solid wire and the liquid droplet. The arc is re-established and the current decreases. The representative current point moves to point 4 from point 2 through point 3. Controlling the current and voltage in this manner reduces spattering to a minimum.

When the current decreases, the power source characteristic is converted to PSC3. The representative point moves to point 5 from point 4 along the down side of PSC3 and arrives at the equilibrium point 5, determined by the wire feed rate and the power source characteristic. Since the arc occurs from the under side of the droplet in a $CO_2$ atmosphere, the droplet in the high arc current period is lifted by the arc pressure. When the power source characteristic is converted to PSC1 after a half period of arc, the representative point moves to point 6 from point 5. Since the current decreases and the arc pressure also decreases, the droplet moves towards the weld pool. At the same time, the surface of the weld pool moves to the direction of the droplet, and the droplet comes in contact with the surface of the weld pool.

Therefore, in order to stabilize metal transfer, the proposed power source is used to make the droplet oscillate and assist transfer to the weld pool. When the resistance of the interface between the solid wire and the liquid pool reduces, the current surges. The arc is established in the low current region. The concepts for the conversion of the power source characteristic and for $CO_2$ short-arc welding are illustrated in Figure 15.6.

## 15.4 AUTOMATIC SETTING OF WELDING PARAMETERS BY FUZZY INFERENCE

In $CO_2$ short-arc welding, it is important that the weld pool width and the cooling time are kept constant so as to stabilize the weld pool shape and to obtain a good quality welding result. Moreover, spatter should be minimized. However, welding at high speed is desired to obtain high efficiency. The specification (cooling time and pool width) for the weld is determined from the thickness and the material of the plate. Typical relationships between arc current, short-circuit current and welding speed in combination with cooling time $CT$ and pool width $W$ are shown in Figures 15.7, 15.8 and 15.9. In Figure 15.7, the welding speed is constant (80 cm/min). In Figure 15.8, the short-circuit current is constant (300 A). In Figure 15.9, the arc current is constant (250 A). From Figure 15.9, the short-circuit current required to maintain the width increases when the welding speed is increased. The amount of spatter is proportional to the short current.

The relationship between spattering and efficiency is complex, therefore. Skilled welders can decide the optimum welding parameters from their relationships, satisfying the specification and considering the amount of spatter and the efficiency. The process adopted by skilled welders to determine the welding parameters is expressed in IF–THEN form. In the new control system the method for optimizing the welding parameters uses fuzzy inference. The inference rule is constructed from the relationship between the cooling time and the width.

First, the rule for obtaining the arc current and the short-circuit current is constructed from Figure 15.7. Let the reference of the cooling tim $RCT$ and the pool width $RW$ be respectively 0.5 min and 7 mm at point 9 in Figure 15.7. When the present arc current and the present short-circuit current are

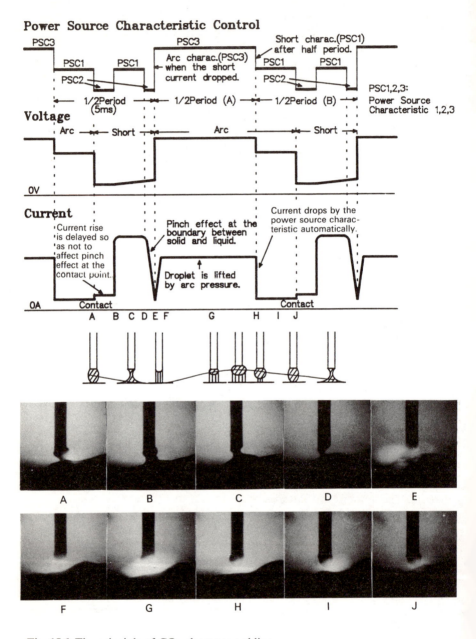

**Fig. 15.6** The principle of $CO_2$ short-arc welding.

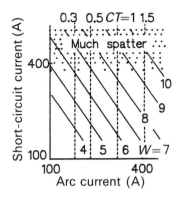

**Fig. 15.7** Relationship between arc current and short-circuit current.

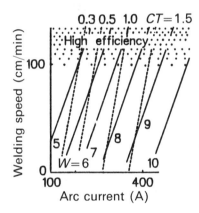

**Fig. 15.8** Relationship between arc current and welding speed.

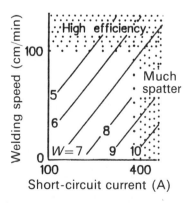

**Fig. 15.9** Relationship between short-circuit current and welding speed.

at point 2, the width is narrower than that of the reference and the cooling time is longer than that of the reference. In this case, the reference values are obtained by which the arc current is made to increase while the short-circuit current is made to decrease. This is rewritten in IF–THEN form as follows:

$$\text{IF } e_{CR} \text{ is P and } e_W \text{ is N then } \Delta AC \text{ is N and } \Delta SC \text{ is N.} \qquad (1)$$

where P and N are the fuzzy variables shown in Figure 15.10(a). This means that if the deviation $e_{CT}$ ($= CT - RCT$) of the cooling time is positive (P) and the deviation $e_W$ ($= W - RW$) of the pool width is negative (N) then the variaton of the arc current $\Delta AC$ and of the short-circuit current $\Delta SC$ are negative (N). The arc current and the short-circuit current are obtained from summing up $\Delta AC$ and $\Delta SC$, respectively. In Figure 15.10(a), $e$ corresponds to $e_W$, $e_{CT}$, $\Delta AC$, and $\Delta SC$.

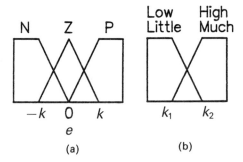

**Fig. 15.10** The fuzzy variables: (a) deviations of width and cooling time, variations of arc current and short-circuit current; (b) efficiency (welding speed) and amount of spatter (short-circuit current).

The rule for the arc current and the short-circuit current is constructed in terms of the typical points 1–9 in Figure 15.7 and is listed in Figure 15.11. Though this rule is for constant welding speed, it can be assumed that the relationships between the parameters are maintained regardless of the change of the welding speed.

|  |  | Cooling time | | |
|---|---|---|---|---|
|  |  | N | Z | P |
| **Width** | P | P:N | Z:N | N:Z |
|  | Z | P:N | Z:Z | N:P |
|  | N | P:Z | Z:P | N:P |

Arc : Short

**Fig. 15.11** The rule for cooling time, width, arc current and short-circuit current: P, positive; Z, zero; N, negative.

Next, the welding speed is determined from the efficiency and the amount of spatter. The fuzzy variables for spatter and efficiency are constructed from two levels of fuzzy variable, which are 'much' or 'little' for the amount of spatter and 'high' or 'low' for efficiency, as shown in Figure 15.10(b). The rule for avoiding spatter is shown in Figure 15.12. The optimum welding parameters are obtained from Figures 15.7, 15.8, 15.9, 15.11 and 15.12. For example, when $RCT$ and $RW$ are respectively 0.5 min and 7 mm, the welding parameters arc current, short-circuit current and welding speed, determined by fuzzy inference, are 390 A, 250 A and 98 cm/min respectively. In Figure 15.10, $k = 1$ mm for $e_W$, $k = 1$ min for $e_{CT}$, $k = 5$ A for $\Delta AC$, $k = 5$ A for $\Delta SC$, $k = 5$ A for $\Delta AC$, $k_1 = 90$ cm/min and $k_2$ for the efficiency (corresponding to the welding speed), and $k_1 = 390$ cm/min and $k_2 = 410$ A for the spatter amount (corresponding to the short-circuit current).)

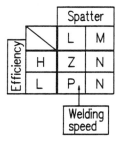

**Fig. 15.12** The rule for spatter, efficiency and welding speed: M, much; L, little; H, high; L, low.

## 15.5  CONCLUSION

The weld pool in $CO_2$ short-arc welding has been observed during the period of short circuiting when the arc is extinguished by using a CCD shutter camera. A variable power source characteristic is proposed for improving the stability and self-regulation of metal transfer in $CO_2$ short-arc welding. It has been possible to obtain short-arc welding with the absence of spatter. The wire melting and base metal melting have been controlled by adjusting the arc and short-circuit characteristics. The optimum welding parameters are obtained by fuzzy inference.

## REFERENCES

1. Ohshima, K., Mori, Y., Kaneko, Y., Mimura, A., Kumazawa, T., Kubota, T. and Yamane, S. (1990) Image processing and control of weld pool in short-arc welding, in *Proceedings Fifth International Symposium of the Japan Welding Society*, pp. 483–98.

2. Ohshima, K., Morita, M., Fujii, K., Yamamoto, M. and Kubota, T. (1987) *Journal of Japan Welding Society*, **5**, 304.
3. Ohshima, K., Mori, Y., Ma, P. and Yamane, S. (1991) Sensing and control of $CO_2$ short arc welding, in *Proceedings International Conference on New Advances in Welding and Allied Processes*, Benjin, China, pp. 101–105.

# CHAPTER 16

# Application of a fuzzy neural network to welding line tracking

*Kenji Ohshima, Satoshi Yamane, Yasuchika Mori*
*and Peijun Ma*

This chapter deals with the problem of tracking the welding line in an arm-type welding robot. The kinematic equations of the robot are complex, since the moment of inertia changes with changes in torch position as the welding robot is operated. A method based on fuzzy control is therefore proposed for tracking the welding line. The fuzzy variable and production rules play an important role in fuzzy control. In this chapter, the fuzzy variable and the production rules are decided by expert knowledge; the values of the fuzzy variables are decided by experiment. Moreover, in order to improve the tracking performance, a neural network is used to recognize the form of the welding line and the result is used to adjust the fuzzy variable. The validity of the proposed tracking method is verified by the experiments.

## 16.1 INTRODUCTION

This chapter deals with the problem of tracking the welding line in an arm-type welding robot. The state equations which describe the motion of the joint are complex, since the moment of inertia changes with changes in torch position as the robot is operated. It may be difficult to trace the welding line by methods based on modern control theory.

A fuzzy control method is therefore proposed for tracing the welding line. It is important to decide the correct fuzzy variables and rules in fuzzy control. In the work described in this chapter, the fuzzy variables and rules are decided from the knowledge and experience of robot operators without

*Sensors and Control Systems in Arc Welding.* Edited by Hirokazu Nomura.
Published in 1994 by Chapman & Hall, London. ISBN 0 412 47490 5

the motion model of the robot; the rules are determined from the relationship between the advancing direction of the torch and the position 3 mm ahead of the torch. Tracking experiments are performed based on the deviation at a position 3 mm ahead of the torch. Tracking is delayed when the welding line bends by 60°. If the angle of the welding line is obtained, the fuzzy variables can be adjusted according to the angle. The tracking performance is improved by adjusting the fuzzy variables. The angle may be formed by recognizing the bend pattern of the welding line.

To improve the tracking performance, a neural network is used to recognize the angle of the welding line. The tracking result with the neural network is better than that without it. The validity of the proposed method is verified by tracking experiments.

## 16.2   SYSTEM FOR TRACKING THE WELDING LINE

The tracking system is shown in Figure 16.1. The image of the plane including the welding line is taken into the image memory with a CCD shutter camera. A computer processes the image and calculates the distance (the deviation) between the torch position and the welding line. The targeted torch position is inferred from the deviation and its variation. The number of stepper motor pulses required to move the robot axes to the targeted torch position is calculated and appropriate control signals are sent from the robot controller.

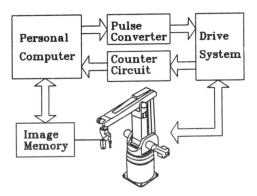

**Fig. 16.1** The tracking system in a multijoint system.

## 16.3   FUZZY CONTROLLER

The fuzzy controller for the multijoint robot is shown in Figure 16.2. The deviation *e*, which is the distance between the welding line and the position 3 mm ahead of the torch in the direction of torch motion and the variation

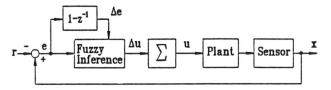

**Fig. 16.2** The fuzzy controller for tracking the welding line.

$\Delta e$ of the deviation are adopted as the input variables of the fuzzy controller. The variation $\Delta u$ of the targeted advancing direction of the torch is adopted as the manipulating variable. The relationship between the torch position and the fuzzy variables is shown in Figure 16.3. In the fuzzy control method, the manipulating variable is inferred both from the rules and from the input variables. In this chapter, the rules listed in Figure 16.4 are designed from the knowledge and the experience of the operator and the fuzzy variables shown in Figure 16.5 are determined from experiment.

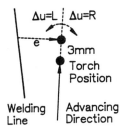

**Fig. 16.3** The relationship between the torch position and the fuzzy variables.

| $\Delta e$ \ $e$ | L | Z | P |
|---|---|---|---|
| R | Z | L | L |
| Z | R | Z | L |
| L | R | R | Z |

**Fig. 16.4** The rules for tracking: R, right; Z, zero; L, left.

## 16.4  NEURAL NETWORK

If the angle of the welding line is given, the tracking precision may be improved by adjusting the parameter $kdu$ of the fuzzy variable. A three-layer feed-forward network is therefore constructed for classification of the angle. The angle of the welding line is classified in five divisions: straight, slight left, left, slight right, and right. The distance between the reference line and the welding line, as shown in Figure 16.6, is used for information about the

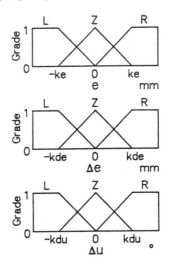

**Fig. 16.5** The fuzzy variables.

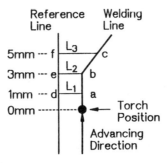

**Fig. 16.6** The relationship between the reference line and the welding line.

angle by which the welding line is bent. $L_1$, $L_2$ and $L_3$ are the distances between the reference line and the positions $a$, $b$ and $c$, respectively; $d$, $e$ and $f$ are 1, 3 and 5 mm from the current torch position. Hence, the number of cells in the input layer is 3 and the number of cells in the output layer is 5. The neural network is shown in Figure 16.7. The backpropagation method is used to train the network. In learning, tutor patterns for preset input patterns are required. Here, the learning is performed for various cases in which the welding line is bent by 30° and 60°. Twelve cells are needed in the hidden layer so that the error in learning converges to 0.

## 16.5  TRACKING EXPERIMENTS

The torch is moved to the start point of the welding line. The deviation $e$ and its variation $\Delta e$ are calculated by the computer from information

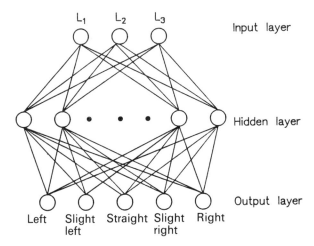

**Fig. 16.7** A three-layer neural network for recognizing the angle of the welding line.

received from the sensor. The variation $\Delta u$ of the targeted advancing direction (the manipulator variable) is derived by fuzzy inference. The angle of each joint is calculated from the targeted advancing direction by solving the inverse kinematic problem in order to move the torch to the targeted position. The number of pulses is calculated from the angle. $ke$ and $kde$ are determined from the various tracking experiments. The reference line is drawn in the same direction as the torch advancing direction. From the distance between the welding line and the reference line, the angle of the welding line is classified by the neural network. The magnitude $kdu$ of the fuzzy variable is adjusted according to the classification of the angle; $kdu$ for straight, for slight left or right, and for left or right, is 20, 100, and 60 respectively.

The tracking experiment for the welding line shown by the thin line in Figure 16.8 was first performed without the neural network, with fuzzy

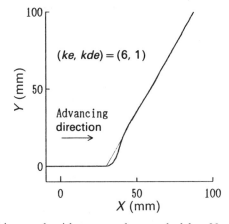

**Fig. 16.8** The tracking result without neural network: $kdu = 20$.

variables (*ke, kde, kdu*)=(6, 1, 20). The welding line bent by 60°. The tracking results are denoted by the thick line. Overshoot can be seen in Figure 16.8, showing lack of responsiveness to such a large change in direction of the welding line. In order to decrease the overshoot, *kdu* was increased and hence the gain of the controller increased. The tracking result is shown in Figure 16.9. The overshoot has been eliminated, but the trajectory has oscillated about the welding line. Next, the tracking experiment was performed with the neural network. The tracking result is shown in Figure 16.10; good performance was obtained, the steady-state error and the overshoot being zero.

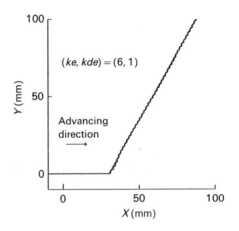

**Fig. 16.9** The tracking result without neural network: *kdu*=60.

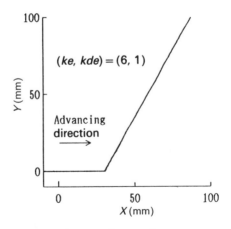

**Fig. 16.10** The tracing result with neural network.

## 16.6 CONCLUSION

This chapter has dealt with the problem of tracking the welding line in an arm-type welding robot. Tracking the welding line is difficult using methods based on modern control theory, and a fuzzy control method was therefore used. The fuzzy variable and the production rule play an important role in fuzzy control. In the work described in this chapter, they were decided from the knowledge and the experience of the expert. The fuzzy variable was adjusted by using a neural network to recognize the form of the welding line. The validity of the proposed tracking method was verified by various experiments.

## REFERENCES

1. Ohshima, K., Mori, Y., Kaneko, Y., Mimura, A., Kumazawa, T., Kubota, T. and Yamane, S. (1990) Groove gap detection and weld pool control using CCD camera, in *Proceedings Fifth International Symposium of the Japan Welding Society*, Tokyo, Japan, April, pp. 489–94.
2. Yamane, S., Kitahara, N. and Ohshima, K. (1990) Application of fuzzy adaptive control to a welding robot, in *Proceedings Second International Symposium on Signal Processing and its Applications*, Gold Coast, Australia, August, pp. 271–4.
3. Ohshima, K., Mori, Y., Ma, P. and Yamane, S. (1991) Fuzzy control in tracking of welding line, in *Proceedings International Conference on New Advances in Welding and Allied Processes*, Benjin, China, May, pp. 138–41.

# CHAPTER 17

# Magnetic control of the arc in high-speed TIG welding

*Shizuo Ukita, Taisuke Akamatsu*
*and Kan'ichiro Shimizu*

When the welding speed in TIG welding is kept high, the arc becomes unstable and tends to extinguish. By applying a magnetic force to arc, the directionality of the arc can be controlled so as to flow along the extension line of the electrode axis at the shortest distance to the base metal. To detect the direction of arc flow, the arc light is observed via glass fibre optics positioned at fixed points around the electrode and transmitted to photosensors. The light is converted to electrical signals which are fed into a computer. Subsequently, the light intensity at respective points, and their positions, are calculated. These procedures are programmed in the computer, and provide automatic control of the arc position by electromagnetic deflection of the arc to maintain uniform light distribution around the control electrode.

## 17.1 INTRODUCTION

Recently the demand for aluminium sheets has grown, and attention has been paid to the feasibility of welding thin sheets, increase in welding speed, automation and so on. Butt welding of thin aluminium sheets (thinner than 0.3 mm) by TIG welding becomes very difficult because of the high heat conductivity and high thermal expansion coefficient of aluminium.

The distortion of aluminium sheet during welding causes misalignment of the butted edges. When the welding speed is increased, this misalignment diminishes, but the arc is deflected backwards and becomes unstable or may extinguish.

*Sensors and Control Systems in Arc Welding.* Edited by Hirokazu Nomura.
Published in 1994 by Chapman & Hall, London. ISBN 0 412 47490 5

The authors have succeeded in correcting the deviation of the arc to enable high-speed welding by various methods [1][2].

This chapter describes one of these techniques which uses a microcomputer to detect the directionality of the arc using sensors, and correct and control the flow of the arc using magnetic force.

## 17.2 CONSTRUCTION OF THE SYSTEM

Figure 17.1 shows a TIG arc viewed coaxially from above. The centre O is on the extension line of the electrode axis, and the N–S poles of an electromagnet are installed opposite each other perpendicular to the weld line YY'. The arc flows from the electrode to the base metal.

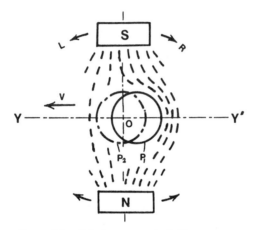

**Fig. 17.1** Arc flow directed forcibly by magnetic field.

As the welding speed $V$ is increased, the arc root on the base metal shifts towards the Y' side deviating from the centre O. When d.c. current is applied to the electromagnet to generate a magnetic field between the N–S poles in Figure 17.1, according to Fleming's rule, the arc root is pushed back from position $P_1$ to the original position $P_2$. In this case, if the intensity of the electric field is increased further, it is possible to shift the arc beyond O to the Y direction.

When the arc deviates from the YY' line, by turning the N–S poles around O, a magnetic field can be generated deflecting the arc in directions across the welding line and restoring its position central on the weld line. How the direction of arc flow changes with the level of magnetic flux density of a given magnetic field has already been reported in [2].

In order to detect the position of the arc when it is shifted, glass fibres and photosensors are used. Image sensors used for CCD cameras are light-receiving elements which detect the quantity of light in photosensors.

Photodiodes or phototransistors are used; the fast response PIN type is desirable, but the sensitivity is not important.

Figure 17.2 shows the experimental set-up. The stepper motor and the electromagnet are attached to a TIG torch, and the electromagnet can be rotated around the axis of the torch centre by the stepper motor through gears 1 and 2.

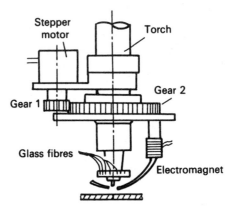

**Fig. 17.2** Schematic diagram of sensing and controlling apparatus.

Holes 0.6 mm in diameter are drilled, in the disk attached to the tip of the torch, and glass fibres 0.5 mm in diameter are inserted.

The tips of the glass fibres are directed to the base metal surface and are arranged as shown in Figure 17.3 to receive the arc light, which is converted to electromotive force by the phototransistor shown in Figure 17.4; its intensity is adjusted by the input controller and stored in the memory of the computer by A/D conversion. After the data for all the points have been read in by a multiplexer, the position of the highest value among them is calculated by the computer. The stepper motor is operated to turn the magnetic poles by the angle $\theta$ from the present position to that point.

For example, when that position is 18° clockwise from the present position (assuming a gear ratio of 1:4 and 1.8° per step), the number of pulses given to the motor results in 40 turns to the right. Also, according to whether the radial position is on the $L_1$ row close to the centre or on the $L_3$ row, the intensity of current applied to the magnet is decided and the switch of an electromagnetic relay corresponding to it is selected and turned on.

Immediately after that, the multiplexer functions to return to the first procedure of taking in the arc light and then repeating the procedures, correcting the flow of arc until welding is finished. Arc light is not taken directly into photodiodes or phototransistors because it is too intense and contains some noise. Also, the high-frequency wave at the time of starting the arc intrudes into the diode circuit, and the photodiodes are too large for use as sensors.

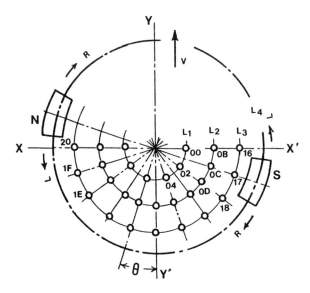

**Fig. 17.3** Position of optical fibre sensor.

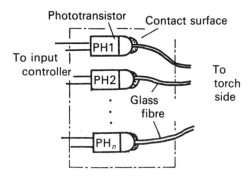

**Fig. 17.4** Contact surface of phototransistor and optical fibre.

Optical fibres have many merits, however. Electrical insulation is easy; the dimensions are small; a multipoint arrangement is possible, and noise does not invade the measurement system.

Figure 17.5 shows the photocoupler used for cutting off noise; signals passed through the photocoupler are amplified by the Darlington circuit and sent to the relay. In this way, electrical conduction between A and B is zero, and the intrusion of noise as well as high-frequency waves is prevented.

Figure 17.6 shows the block diagram of the system.

It was found that when the programming was written in a high-level language, and run by assembling to machine language, the time required for reading the data for all points by means of the photosensor and finishing the comparison calculation was about 7.2 ms.

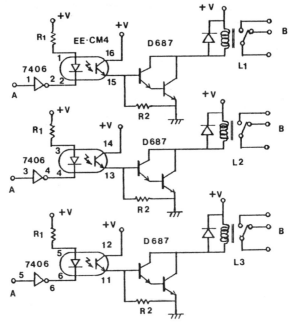

**Fig. 17.5** Photocoupler used for cutting off noise.

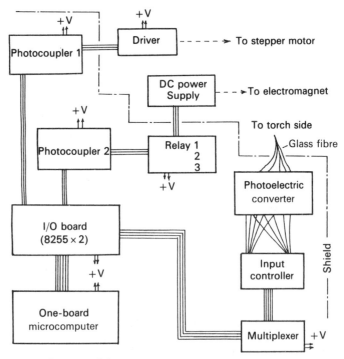

**Fig. 17.6** Block diagram of the system.

An experiment for butt welding of aluminium sheets of 0.3 mm thickness at a welding speed of 24 m/min was carried out by using the system, and experiments continue.

## 17.3  CONCLUSION

In order to realize the high-speed welding of thin aluminium sheets, a method has been developed for controlling the directionality of the arc with a microcomputer using glass fibres, photosensors and electromagnetic force.

This method seems to be applicable not only to high-speed TIG welding but also to low welding speed, to provide correction of arc flow, control of weaving and arc rotation by changing the computer program.

## REFERENCES

1. Ukita, S., Akamatu, T. and Shimizu, K. (1985). Welding condition of thin A1 sheet of high travel speed DCSP TIG welding, in *Proceedings National Meeting of the Japan Welding Society*, **36**, p. 156.
2. Ukita, S., Akamatu, T. and Shimizu, K. (1986). High speed TIG welding of very thin A1 sheets with electro-magnetic field, in *Proceedings National Meeting of the Japan Welding Society*, **39**, p. 284.
3. Arata, Y. and Maruo, H. (1971). Magnetic control of plasma arc and its application for welding, IIW Document IV 53–71.

# CHAPTER 18

# Adaptive control of welding conditions using visual sensing

*Ken Fujita and Takashi Ishide*

Multilayer welding of thick steel plate is still, essentially, performed manually, although many fabricators would like to introduce adaptive control of welding conditions aimed at continuous automatic welding of all layers in multipass multilayer GMA welding.

## 18.1  INTRODUCTION

Various welding sensors have recently been developed for practical application to automatic welding operations. However, multilayer welding of thick steel plate by multiple passes is still performed manually, despite much research in this area.

This chapter describes a GMA welding visual sensor, utilizing a CCD camera together with a rotary mirror-type displacement sensor. It senses the welding situation in real time in order to feed it back and control the welding conditions so as to realize welding under optimum conditions.

## 18.2  CONSTRUCTION

Figure 18.1 shows configuration of the system, which comprises: the torch controller that realizes the overall control; the visual sensor composed of an interference filter and a CCD camera; the device for processing images of the joint or previous weld run sent from the visual sensor; the laser displacement sensor for detecting the bead's sectional form inside the groove in front of the welding; the processing unit for the laser displacement sensor; and the welder.

*Sensors and Control Systems in Arc Welding.* Edited by Hirokazu Nomura. Published in 1994 by Chapman & Hall, London. ISBN 0 412 47490 5

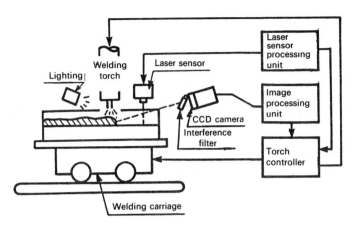

**Fig. 18.1** System configuration.

### 18.2.1 Visual sensor

A small charge-coupled device (CCD) camera was used for the visual sensor. This was adopted through evaluation of compactness, excellent durability, and high applicability to welding [1].

Optical filtering by means of an interference filter and infrared filter with visible range of 500–700 nm was used so that remote monitoring under the high-intensity MAG arc light, could be realized.

**Fig. 18.2** Sensing situation.

### 18.2.2 Rotary mirror laser displacement sensor

This sensor is constructed as shown in Figure 18.3. It comprises the laser diode (830 nm) and the light detector, and makes measurements based on the principle of trigonometric triangulation with the PSD line sensor. In

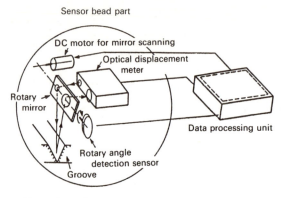

**Fig. 18.3** Composition of welding groove profiling sensor.

order to detect the section form of the groove, the mirror is rotated and the inside of the groove is traversed. The $R-\theta$ coordinates are converted into $X-Y$ coordinates as shown in Figure 18.4.

This sensor has a measurement range of $\pm 20$ mm, and its detection is positional 100 mm before the centre of the arc, and about 40 mm above the groove surface; the scanning speed is 100 rev/min. The rotary mirror was employed in order to reduce the sensor dimensions, and also to allow high-speed scanning.

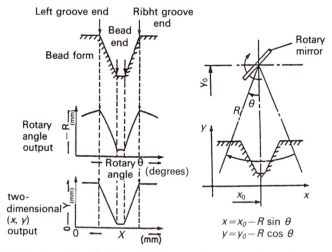

**Fig. 18.4** Method used by the laser displacement sensor to detect forms.

### 18.2.3  Automatic welding control

Welding was realized automatically by carrying out the layering process shown in Figure 18.5. The main elements of the control system are explained below.

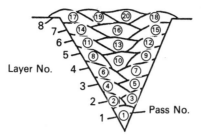

**Fig. 18.5** Multipass layering process.

### (a)  Image recognition method

The image captured by the visual sensor is taken into the frame store ($512 \times 512 \times 8$ bits $= 256$ kbytes) and the images of left and right edges of the groove, the lower end of the wire and the lower end of the torch (Figure 18.6) are processed. Thus the torch position in relation to the groove is recognized and this location data is communicated to the torch controller in order to realize torch position compensation in real time.

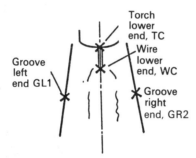

**Fig. 18.6** Image information.

Various automatic recognition methods were simulated from the characteristic elements of the light density distribution by focusing on the various types of information shown in Figure 18.7 and feature recognition was realized by the methods shown.

### (b)  Groove profile detection method

The groove sectional profile was obtained from the laser displacement sensor's output. The bead joint position and the sectional area are computed. The form thus drawn and the recognition methods are shown in Figure 18.8.

### (c)  Torch position control

The torch position was computed by using the bead joint position as shown in Figure 18.9. For the layering procedure, the torch target position was

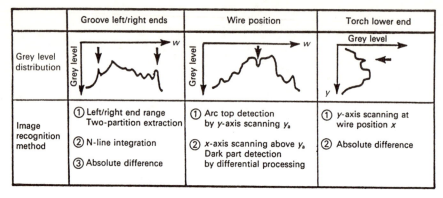

| | Groove left/right ends | Wire position | Torch lower end |
|---|---|---|---|
| Grey level distribution | (grey level graph) | (grey level graph) | (grey level graph) |
| Image recognition method | ① Left/right end range Two-partition extraction ② N-line integration ③ Absolute difference | ① Arc top detection by y-axis scanning $y_a$ ② x-axis scanning above $y_a$ Dark part detection by differential processing | ① y-axis scanning at wire position x ② Absolute difference |

**Fig. 18.7** Image recognition method.

| | Groove left/right ends | Joint position of overlying bead | Both sides overlying | Section area |
|---|---|---|---|---|
| Bead form | 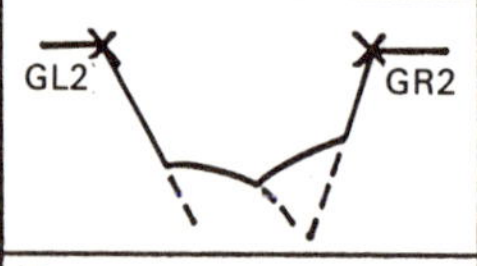 GL2 ... GR2 | 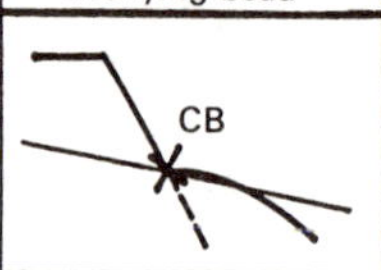 CB | 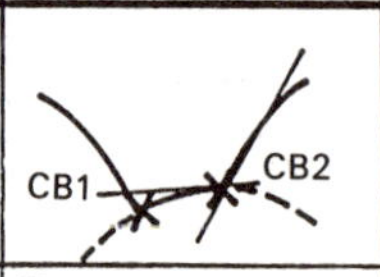 CB1 ... CB2 | 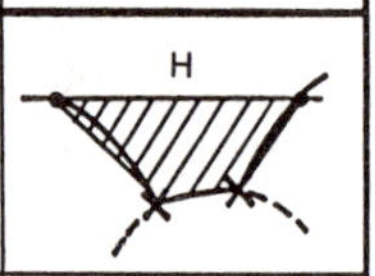 H |
| Recognition method | ① Averaging left and right N points ② Searching point with difference of δ or more ③ Approximate linearization of groove slope ④ Minimum difference point of groove angle | ① Max tilt angle difference of approximate line of groove slope | ① Searching minimum y point ② Max tilt angle difference of approximate line of groove slope | ① Trapezoidal computation of the specified overlaying height and the two points shown to the left |

**Fig. 18.8** Groove sectional profile and recognition method.

| | First pass | Second pass | Third pass | Tenth pass |
|---|---|---|---|---|
| Torch target position for each pass | 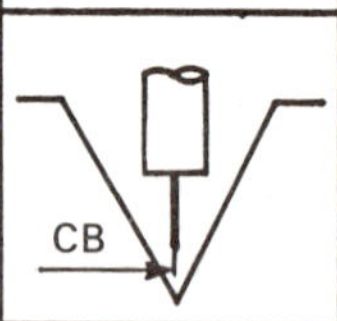 CB | 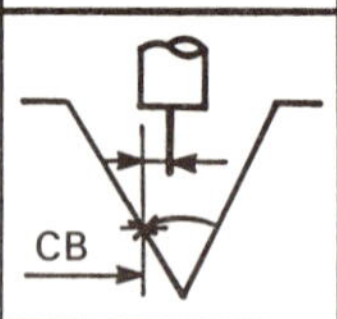 CB | 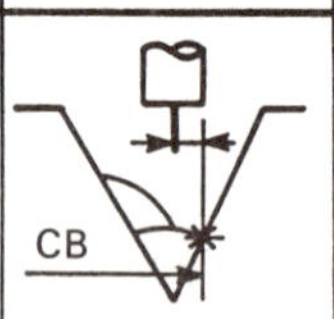 CB | 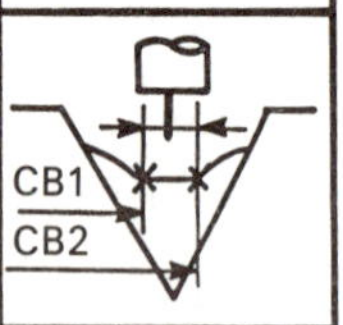 CB1 CB2 |
| | XT = CB | CB + m | CB − m | (CB1 + CB2)/2 |
| Deviation of present torch position from groove end | $XL = (WC - GL1)/(GR1\text{-}GL1) \times (GR2 - GL2)$  $\Delta x = XL - (XT - GL2)$ | | | 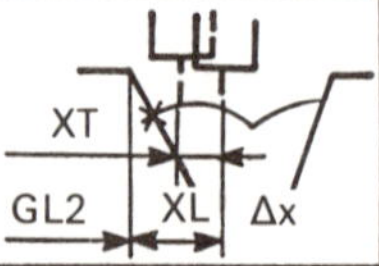 (diagram: XT, GL2, XL, Δx) |

**Fig. 18.9** Torch target position and deviation detection method.

computed by setting the reference point as follows. For the first pass, the centre of the groove; the point on the left of the weld run for passes 2, 4, 6, 8, 11, 14, 17 and 18; point on the right of the weld run for passes 3, 5, 7, 9, 12,

15 and 19 and the centre between the two immediately previous passes for overlaying the rest of the passes.

For the torch position, the difference from the current torch position, which was obtained through image recognition, was computed in order to control the magnitude of displacement.

As there is a displacement of 100 mm between the sensing position and the welding position, the control was delayed by a corresponding amount to match this displacement.

### (d)  *Welding speed control*

The welding speed is controlled so that the volume of welded overlay forms a suitable build-up. For passes 3, 5, 7, 19, 13, 16 and 20 in the layering procedure, the sectional area of overlay was sought as shown in Figure 18.8, and the carriage's speed was regulated from the empirical relationship between the sectional area $A$ and the welding speed $V$. For the other passes, control was exercised so that the overlay would reach the specified height. For this too, the speed was controlled in accordance with the empirical relationship between the overlay height $H$ and the welding speed $V$.

## 18.3  RESULTS

### 18.3.1  Image recognition

The spectrum analysis of the MAG welding arc light with and without the filter is shown in Figure 18.10. The effects of the filter are clearly demonstrated.

For image processing, about 200 images were processed from the first layer to the final layer; when the error was $\pm 1.0$ mm, the groove recognition rate was about 75%, and the wire recognition rate was about 65% (Figures 18.11 and 18.12). By comparing these with the preprocessing displays and by

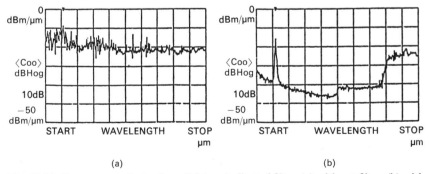

(a)  (b)

**Fig. 18.10** Spectrum analysis of arc light and effect of filter: (a) without filter; (b) with filter.

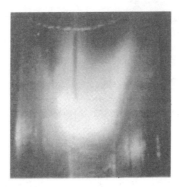

**Fig. 18.11** Monitoring image.

**Fig. 18.12** Image-processing picture.

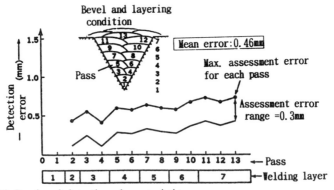

**Fig. 18.13** Bead-end detection characteristics.

narrowing the masking range for the recognition position, automatic control was made possible.

### 18.3.2 Recognition by laser displacement sensor

For the recognition of bead joint position 20 points were studied for each of the 13 passes and the error thus found was ±0.3 mm (Figure 8.13).

For the area, judging the bead height after beads were placed for seven object passes, the control remained within the tolerance range of about ± 1.5 mm.

## 18.4   CONCLUSION

We have described an example of a multilaser automatic welding system developed by our company. Automation in welding should make further development in the future, and we should like to achieve higher precision, faster processing and lower cost for this system, to make it more widely available and applicable.

## REFERENCE

1. Development of Visual Sensor for Gas Shielded Arc Welding, Mitsubishi Heavy Industries Ltd, Japan.

# CHAPTER 19

# Automatic welding system with laser optical sensor for heavy-walled structures

*Hirokazu Wada, Yukio Manabe and Shigeo Inoue*

In the welding of large, heavy-walled steel structures such as pressure vessels, the main need is for labour saving and uniform quality of welds. Ageing of welders and a shortage of skilled workers also present problems. To solve these problems, automatic welding has been employed in many fields. Automatic, unattended welding requires sensor systems suitable for each object to be welded. This chapter describes a sensing technique for determining the cross-sectional profiles of grooves, using a laser optical sensor for automatic multipass welding and an intelligent automatic welding system incorporating this technique.

## 19.1 SENSOR SYSTEM AND SENSING TECHNIQUE

### 19.1.1 Laser optical sensor

Figure 19.1 shows the fundamental construction of the laser optical sensor. A laser beam emitted from the beam source is focused into a spot beam and on to the object to be welded. Part of the beam reflected from the object surface is sensed by a photoelectric element through a filter and condensing lens to determine the distance between the sensor and object surface by means of triangulation.

The sensor can therefore detect a groove or bead shape when it scans across them. When an appropriate filter is selected for the reflected beam,

*Sensors and Control Systems in Arc Welding.* Edited by Hirokazu Nomura. Published in 1994 by Chapman & Hall, London. ISBN 0 412 47490 5

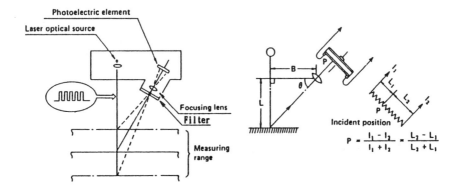

**Fig. 19.1** Fundamental construction of the laser optical sensor, and principle of distance measurement.

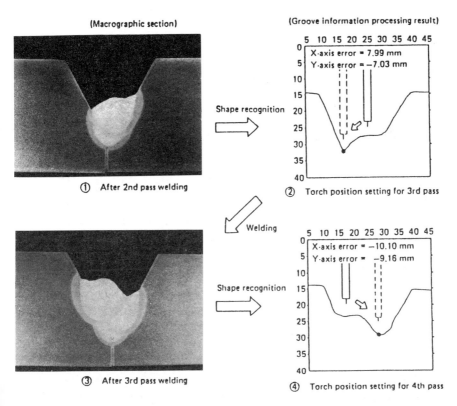

**Fig. 19.2** Torch position automatic control test: result for multipass submerged arc welding.

external disturbance can be reduced, and measuring errors are less than 0.5 mm.

The sensor is fully adaptable as a weld line tracking sensor.

### 19.1.2   Scanning position control technique

A basic study was made of a weld line tracking technique to control the torch position, using the groove information obtained by the sensor and processed by microcomputer.

A program was prepared which would aim the torch at the deepest spot in a joint, and a basic welding test was performed using this program. Examples of test welds are shown in Figure 19.2. The torch aiming position was controlled accurately and good welds were obtained. The applicability of this sensor and basic control method was confirmed.

**Table 19.1** Performance result

| *Item* | | *Measuring accuracy* |
|---|---|---|
| Surface condition | ⎰ Rusted<br>⎨ Painted<br>⎱ Machined | ±0.1 mm |
| Preheating temperature (0–500°C) | | ±0.05 mm |
| Welding arc | | ±0.3 mm |

### 19.1.3   Build-up algorithm and welding condition adaptability control technique

The build-up and tracking algorithm for thick-plate joints is the most important requirement for automated welding. It controls the torch aiming position, taking account of the build-up of previous beads, etc. so that no defects are caused.

Figure 19.3 shows examples of multipass welds. Conventional sensors were unable to detect the grooves and bead shape of these joints, which prevented their automatic welding.

The purpose of the technique proposed here is the detection of datum points (particular points of the groove) for control of the torch position.

These points are located at the intersections of the groove walls or beads as shown in Figure 19.4. The sensor data at these points are differentially processed.

Moreover, using these datum points, a mathematical formula is set up to obtain the aiming position of the torch for each pass, thereby making multipass tracking welding possible.

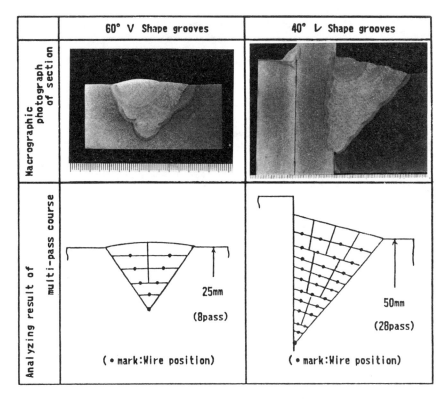

| | 60° V Shape grooves | 40° L Shape grooves |
|---|---|---|
| Macrographic photograph of section | | |
| Analyzing result of multi-pass course | 25mm<br>(8pass)<br>( • mark:Wire position) | 50mm<br>(28pass)<br>( • mark:Wire position) |

**Fig. 19.3** Examples of multipass welds.

Assembly errors and distortion during welding are detected to maintain appropriate welding conditions and produce uniform build-ups. The correction formula is:

$$(A_D/A_n) \cdot a_n = a_D$$

where $A_D$ is the cross-sectional area detected by sensor; $A_n$ is the cross-sectional area of a standard groove (calculated from input data); $a_n$ is the cross-sectional area of a standard weld (calculated from input data); and $a_D$ is the cross-sectional area of the weld after correction. Using this formula, the standard weld cross-sectional area $a_n$ is corrected according to an actual groove, and the required welding conditions (welding current and voltage, speed) corresponding to the calculated weld cross-sectional area $a_D$ are passed to the welding machine so that the quantity of deposited metal and the depth of penetration are controlled adequately for a groove which changes constantly.

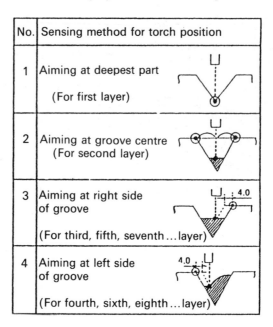

| No. | Sensing method for torch position |
|-----|-----------------------------------|
| 1 | Aiming at deepest part (For first layer) |
| 2 | Aiming at groove centre (For second layer) |
| 3 | Aiming at right side of groove (For third, fifth, seventh...layer) |
| 4 | Aiming at left side of groove (For fourth, sixth, eighth...layer) |

**Fig. 19.4** Method for sensing torch position. ⊙, datum point; ●, torch position. The torch position for each pass is controlled by processing groove shape data from the sensor. The bottom sketch shows an example of the torch-aiming procedure for four-pass welding.

## 19.2  DEVELOPMENT OF PRACTICAL SYSTEM AND APPLICATION RESULTS

Based on the results obtained above, a circumferential seam multipass welding algorithm has been established and a practical system has been developed incorporating it.

Figure 19.5 shows the constitution of a thick-plate automatic welding system.

This system comprises five components. The control systems for the major units were manufactured by us. Each device is equipped with an interface for connection to a microcomputer.

Once the plate thickness is input as an initial value, the system performs multipass welding with no help from the operator, automatically controlling the torch position and welding conditions for each pass.

Figure 19.6 shows an example of application of this system to the welding of circumferential seams in cranes, and Figure 19.7 shows an example of application to chemical equipment. In both cases, automa-

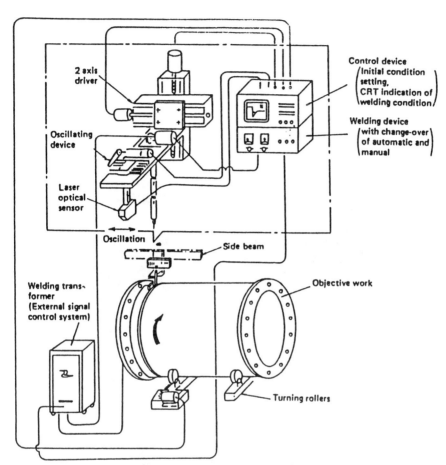

**Fig. 19.5** Construction of the automatic welding system.

**Fig. 19.6** Example of application to welding circumferential seams in cranes.

tion has contributed to stabilizing the quality of welds and greatly reducing the man-hour requirements, thereby having attained the original aims.


### 19.3  CONCLUSION

In the manufacture of large-sized, thick-walled structures, the multipass welding of thick plates is an important process requiring many man-hours. For this welding, we have developed and put into practical use the first fully automatic intelligent welding system ever made in the industry.

(a)

(b)

**Fig. 19.7** Example of application to chemical equipment.

The shape-sensing technique using a laser optical sensor, which is the heart of this system, will have diversified applications in many types of automatic and intelligent welding and working equipment.

## REFERENCES

1. Nomura, Sugitani *et al.* (1981) Automatic profiling method by the use of arc as a sensor (second report), in *Preprints of the National Meeting of the Japan Welding Society*, **28**, 182–3.
2. Matsubara *et al.* (1977) Automation of arc welding by actual article profiling method. *Robot* no. 16.
3. Masumoto, Aratani *et al.* (1983) The present state of sensor and sensor system for arc welding in Japan. *Journal of the Japan Welding Society*, **53**(4), pp. 339–47.
4. Inoue (1984) Arc welding robot sensor. *Welding Technique*, **32**(12), pp. 46–50.
5. Wada, Manabe *et al.* (1984) Study on automatic control technique of submerged arc welding method, in *Preprints of the National Meeting of the Japan Welding Society*, **34**, 64–5.

# CHAPTER 20

# Group-control system for narrow-gap MIG welding

*Hirokazu Wada, Yukio Manabe,*
*Yoshinori Hiromoto and Hideki Yoritaka*

A prerequisite for the automation of narrow-gap arc welding is a sensor which can directly detect various welding phenomena, such as the shapes of arc and molten pool, and torch position. Conventional electromagnetic sensors and arc sensors are not applicable to this task. The authors recognized the advantages of visual sensors and after analysing the arc and molten pool beams detected by such a sensor, have developed an automatic narrow-gap MIG welding system using an image-processing technique.

## 20.1   CAMERA SYSTEM

### 20.1.1   Images

The visual sensor provides information on the light emitted from the arc and the molten pool during welding. These two pieces of information are necessary to control welding conditions.

Figure 20.1 shows an example of spectral measurements of the arc and molten pool light (the broken curves in the figure show values calculated from Planck's radiation law). The peak wavelength of the arc light is in the ultraviolet region and that of the molten pool beam is in the infrared region.

There are considerable differences in the spectral characteristics of the light from the arc and molten pool. To obtain accurate information on these, it is important to set up an optical system which utilizes these differences.

*Sensors and Control Systems in Arc Welding*. Edited by Hirokazu Nomura.
Published in 1994 by Chapman & Hall, London. ISBN 0 412 47490 5

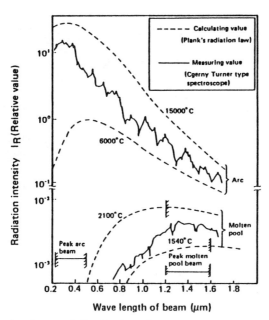

**Fig. 20.1** Example of spectral distribution of arc.

Figure 20.2 shows the characteristics of the image tubes and filters used for the experiment. The sensitivity of a conventional visual vidicon ($Sb_2S_3$) combined with a red filter is mainly located in the visual region as shown by a in the figure. This combination of vidicon and filter is unable to pick up clear images of the arc and molten pool simultaneously with high accuracy. However, a combination of an infrared vidicon (PbS) with an infrared filter has its sensitivity in the infrared region and detects the molten pool selectively, providing good images.

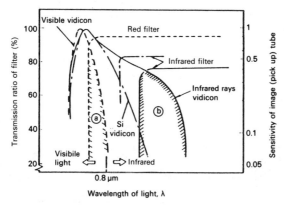

**Fig. 20.2** Selection of image pick-up system.

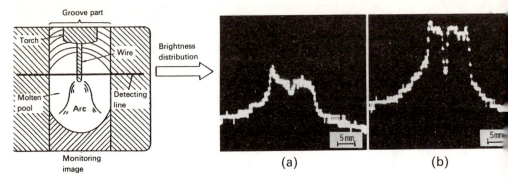

**Fig. 20.3** Results of brightness distribution detection: (a) visible vidicon plus red filter; (b) infrared vidicon plus infrared filter.

Figure 20.3 shows the results of detecting the brightness distribution of a weld. An optical system consisting of a combination ⓑ of infrared vidicon and infrared filter (Figure 20.3(b)) provides an image with clear differences in brightness between the molten pool, left and right groove walls and wire, and can detect the position of the wire.

### 20.1.2   Image processing

A weld was monitored by the image detection system described above, the obtained image information was treated by binary image processing, and the position information necessary for calculation processing was extracted.

Figure 20.4 shows the results of binary image processing of the images of a weld. The value $n$ is the threshold level (relative value) for the binary processing. When $n$ is set at a high level as in Figure 20.4(a), the arc shape can be detected. When the calculation processing is applied to the distribution of

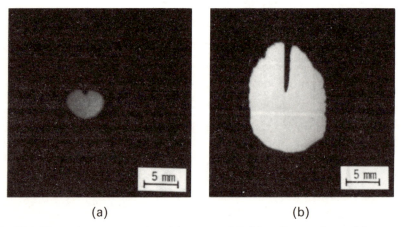

**Fig. 20.4** Binary image processing: (a) arc, $n = 3.0$; (b) molten pool, $n = 1.0$.

the arc shape output signal level along the z axis, the wire extension length and arc length can be detected.

However, when the *n* is set at a low level as in Figure 20.4(b), the molten pool shape can be detected. When the distribution of the output signal level along the *x* axis is extracted and calculation processing is applied to it, the width of the molten pool and the position of the wire can be detected.

From such binary processed images, the signals of wire extension length, wire position, etc. are detected by using the image characteristics mentioned above (Figure 20.5).

If the length and position of the torch across the width are automatically controlled using the information on the extension length and position of wire obtained by the visual sensor, narrow-gap welding can be readily automated, requiring no attendance of the operator. In particular, direct information on the igniting point of the arc allows far more accurate and reliable torch position control than control made by observing the molten pool from above the groove in thick plate.

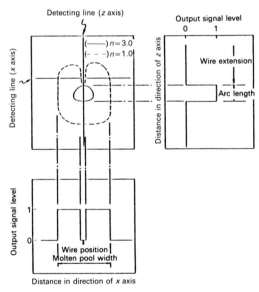

**Fig. 20.5** Image processing for position detection.

## 20.2 APPLICATIONS

Figure 20.6 shows a thick-walled pressure vessel being welded by the automatic welding system developed by us and the weld line obtained.

In the system, images obtained by the visual sensor are subjected to binary processing. From the processed images, the wire extension length, molten pool width (groove width) and wire position are detected. Then these

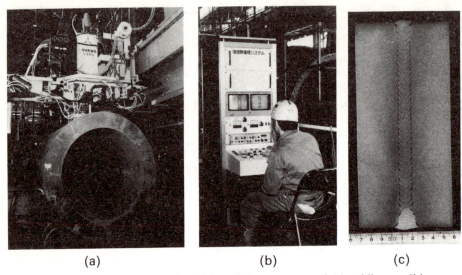

(a)                              (b)                              (c)

**Fig. 20.6** Automatic welding of a thick-walled pressure vessel: (a) welding conditions; (b) operating panel; (c) resulting weld.

items of information are processed by microcomputer to control the torch position.

Once the operator has made the initial setting of the welding torch and has depressed the start button, the system carries out welding automatically without need of any operator in attendance.

## 20.3   CONCLUSION

Based on a visual sensor technique which allows the direct detection of welds by analysing the characteristics of the arc and molten pool light, an automatic welding system for narrow-gap MIG welding has been manufactured and put into practical use.

The developed system produces welds of consistent quality. Moreover, many systems can be operated by one person, greatly reducing the manpower requirement and manufacturing costs.

## REFERENCES

1. Inoue (1981) Image treatment for on-line detection of welding process (second report). *Journal of the Japan Welding Society*, **50**(11), 1118–24.
2. Inoue (1982) Sensor for arc welding robot. *Journal of the Japan Welding Society*, **51**(9), 735–41.
3. Ohmae *et al.* (1984) Study on analysing the characteristics of arc and molten pool beams, in *Preprints of the National Meeting of the Japan Welding Society*, **34**, 66–7.

4. Ohbato *et al.* (1987) Image treatment and digital control of molten pool in pulse MIG welding. *Quarterly Journal of the Japan Welding Society*, **5**(3), 304–11.
5. Nakata *et al.* (1988) Fundamental investigation as to the detection of weld lines and molten pool information, by the simultaneous use of laser beam and narrow-gap interference filter. *Quarterly Journal of the Japan Welding Society*, **6**(1), 123–7.

# CHAPTER 21

# Automatic control technique for narrow-gap GMA welding

*Hiroshi Watanabe, Yoshihide Kondo*
*and Katsunori Inoue*

This chapter describes an automatic control system for narrow-gap welding equipment, developed by BHK. In this system, one operator controls two welding arcs by watching the CRTs. The image-processing system traces the weld groove's centre and keeps the wire extensions to set values, automatically. This type of controller has been applied to actual production. To make welding operations easier, a prototype expert system for the NGW machine with the image processing system has been developed.

## 21.1   INTRODUCTION

Babcock-Hitachi KK (BHK) have developed an original narrow-gap GMA welding (NGW) process, with a wide range of practical applications for pressure vessels, such as those found in thermal power plants, nuclear power plants and chemical plants [1].

In this process, by observing the arc during welding, the operator usually controls many parameters, such as the welding torch position, oscillating width of the wire tip, arc current, and so on. To make the control operation easier, we have developed an automatic NGW machine which has a torch positioning function controlled by microcomputers and image processing [2]. We have also developed a prototype expert system for the NGW machine [3].

*Sensors and Control Systems in Arc Welding.* Edited by Hirokazu Nomura.
Published in 1994 by Chapman & Hall, London. ISBN 0 412 47490 5

## 21.2   PRINCIPLE AND CONSTRUCTION OF THE CONTROL SYSTEM

### 21.2.1   Control parameters of the NGW process

Figure 21.1 shows the BHK NGW process. The process is based on a single-pass one-layer technique using GMA welding. Proper fusion into the groove side walls and the preceding bead is obtained by using a mechanical arc oscillation device. The electrode wire is fed to the molten pool through the contact tube, the feed roller and oscillator. During welding, the electrode is weaved by the bending action of the oscillator combined with the feed rollers. The control parameters of the NGW are: the wire weaving state, the arc state, the melting state in the groove, and trouble shooting.

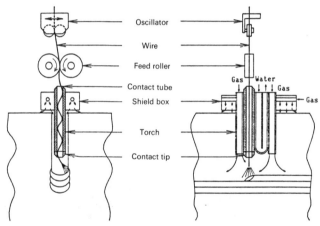

**Fig. 21.1** Schematic of the BHK narrow-gap welding process. Typical welding conditions: electrode diameter, 1.2 mm; current, 280 A; voltage, 28 V; travel speed, 240 mm/min; shielding gas, Ar (80%)+$CO_2$ (20%); groove width, 8–14 mm; base material, low-alloy steel.

### 21.2.2   Image processing

The authors have developed a control system for NGW including an image-processing technique in order to perform NGW automatically and to sense the control parameters mentioned above.

Figure 21.2 shows the image processing flow diagram. A picture of the arc is shown in Figure 21.3, and the measurement values of the image-processing region are shown in Figure 21.4.

The image-processing flow is as follows:

1. The memories, etc., are initialized and the arc image is taken into memories in 64 gradations.

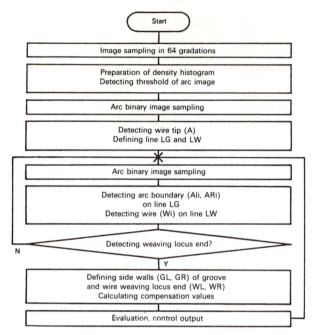

**Fig. 21.2** Image-processing flow diagram.

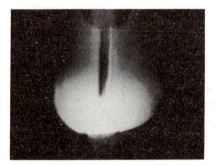

**Fig. 21.3** Typical arc image.

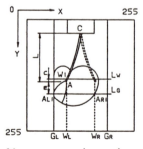

**Fig. 21.4** Measured values of image-processing region.

**Fig. 21.5** Typical binary arc image.

2. The density histogram is repaired, and the threshold for discriminating the arc image is decided.
3. The binary image is sampled (Figure 21.5).
4. The wire tip A is detected. A horizontal line a few levels higher than point A is set as LW (scanning line of the oscillating locus), and a horizontal line a few levels lower than point A is set as LG (scanning line of the arc region).
5. The binary image is sampled.
6. By scaning LG, ALi and ARi (the arc region) are detected. By scanning LW, Wi (the locus of the wire tip oscillated) is detected. The value of the Wi is used to decide whether both ends of the locus have passed or not; if not, the processing steps from (5) are repeated.
7. GL and GR (both side walls of the groove) are decided according to the minimum and maximum values of ALi and ARi, and the groove centre is obtained. By tracing the wire locus Wi, WL and WR (both ends of the oscillated locus) are decided, and the oscillating centre and horizontal correct value are obtained. As the camera is fixed to the torch unit, maintaining the point A at the same position, the wire extension L is always constant. Based on the A position, the correct vertical value is obtained.
8. If the correct values have no errors, the results are transmitted to the welding control system.

The image is sampled 20 to 30 times a second, and this sampling interval is sufficient for practical purposes. Figure 21.6 shows the display of the image-processing result.

By using the image-processing technique, automatic welding seam tracking could be performed for the NGW process.

### 21.2.3 AI control

By controlling the torch positions using the image processing technique, it has been possible to perform fully automatic NGW for a relatively short duration. However, as the welding is continued, the welding conditions vary

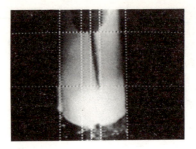

**Fig. 21.6** Result of image processing.

because of such factors as wear of the contact tip of the torch, variation of the wire weaving form, distortion of the groove, and aberration of the rotation work.

To cope with these variations, the operator usually readjusts the relevant parameters, such as the oscillating width and the centre of the wire, the current, the voltage, and the extension. It is difficult to decide which parameter has the most effect when adjusted because the factors are all influenced by each other. The operator judges from his experiences; for example, when the oscillating width is insufficient, expanding it by adjusting the wire-bending mechanism, increasing the wire extension, lowering the arc voltage, changing the wire contact tip, and so on. Selection from these factors is difficult for most welding controllers.

The authors have examined the application of artificial intelligence (AI) control to automate such adjustments, which are usually performed by experienced operators, and have developed a prototype AI controller for NGW.

Figure 21.7 shows a block diagram of the AI controller, which comprises a data analysing system and the expert system. Numerical values and image data are analysed, and the data is transmitted to the expert system as control codes so as to decrease the processing burden of the expert system.

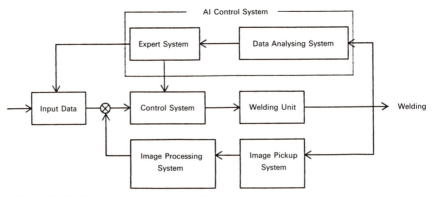

**Fig. 21.7** Block diagram of AI control system.

In the expert system, the welding data is inferred both from the control codes and from the knowledge database which have been elicited from the experience of skilled operators, and from measurement data on NGW. The measurement data and rules are as follows.

#### (a)  *Image data*

In the image-processing system shown in Figure 21.7, two binary images, for the arc and for the molten pool, are sampled, and the groove width, the wire oscillating width, the vertical difference of the wire tip, the arc area, the outline of the arc and outline of the pool are measured or judged.

#### (b)  *Welding current*

The maximum and minimum difference between values of the current correlates the states of the wire oscillating width, the wire extension and the voltage. Variation of the difference correlates one-side wire weaving and deviation of the torch from the groove centre.

#### (c)  *Frequency analysis of welding current*

The peak frequency correlates one-side wire weaving and the deviation of the torch from the groove centre. The peak intensity correlates the wire oscillating width and the wire extension.

#### (d)  *Frequency analysis of the arc sound*

The peak intensity correlates the state of the arc voltage.

Control items in the expert system are the torch location (vertical and transverse directions), the state of the wire oscillation (the width and the centre), the welding voltage, the travelling speed and troubles (the wire oscillation, feeding, power supply, etc.).

Figure 21.8 shows the control flow of the AI system. Every system individually performs data processing, and repeats the processing maintaining its own data. The processed data is transmitted as data codes as required by the AI system.

The expert system infers the state of the control items from each data processing result, and transmits the data codes of the control items to the welding control system when corrective action is required. The welding control system corrects the welding conditions according to the received data codes. When trouble codes are received, the welding is stopped according to the appropriate sequence, and the state of the trouble is displayed.

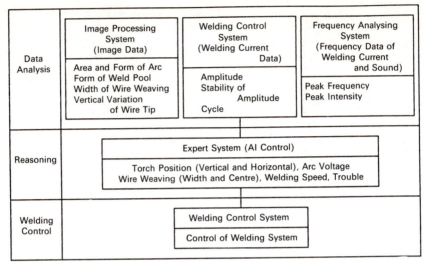

**Fig. 21.8** AI control flow.

## 21.3   APPLICATION

Figure 21.9 shows the construction of the NGW equipment which uses the image-processing control. Two pipes of heavy thickness can be automatically welded at the same time by an operator. Figure 21.10 shows the

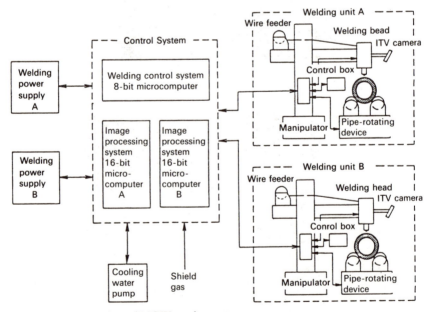

**Fig. 21.9** Construction of NGW equipment.

**Fig. 21.10** Welding controller and welding head.

welding controller with an 8-bit microcomputer for welding control and two 16-bit microcomputers for image processing. The arc picture for the image processing can be taken with a commercial CCD camera through the filter, which consists of heat-absorbing glass and light-shielding glass. The vertical and transverse directions to the weld seam of the welding

**Fig. 21.11** Application of NGW equipment on the manufacturing line.

torch are controlled numerically and can be automatically tracked with the image processing.

The welding equipment has been applied for actual manufacturing lines since 1987. Figure 21.11 shows the application of equipment for the components of thermal power plants which require a high standard of reliability.

## 21.4   CONCLUSION

The authors have developed NGW equipment with both microcomputer controls and image-processing controls, and have applied it to actual manufacturing lines. Furthermore, a prototype welding expert system has been developed to replace the need for the judgement of a human operator.

## REFERENCES

1. Asano, I. (1984) BHK type narrow gap GMA welding process. Narrow gap welding (NGW). Technical Commission on Welding Process, Japan Welding Society, pp. 29–33.
2. Watanabe, H. *et al.* (1987) Automation of narrow gap welding process – application of image processing. Technical Commission on Welding Process, Japan Welding Society, SW-1787-87.
3. Watanabe, H. *et al.* (1988) Automation of narrow gap GMA welding process – applicaton of AI (1). Technical Commission on Arc Physics, Japan Welding Society, 88–693.

# CHAPTER 22

# An automatic control system for one-sided submerged arc welding with flux copper backing

*Nobuyuki Okui*

To determine the correct weld current as a function of the geometrical condition of the groove compensating for deviations from the prescribed standard groove shape, a quantity termed 'joint capacity' is introduced, using which, linear expressions have been established relating the groove geometry to welding current and speed. The empirical relation thus derived gives results agreeing well with the expression given by A.A. Wells relating total heat input to quantity of electrode melting. A laser diode detector system has been devised to detect the position of the groove axis and to discern the groove geometry. The signals from this sensor are fed into an automatic control system incorporating a microcomputer which stores the expressions relating the input from a monitoring sensor to corrective commands to the welding unit.

## 22.1  INTRODUCTION

One-sided submerged-arc welding with flux copper backing is a technique widely used in shipyards. Plates up to 40 mm thick are joined in one pass with double or triple electrodes, and seams of consistently good quality can be produced with high efficiency.

In present practice, however, only the welding conditions applicable to the prescribed standard groove shape are given to the operator. He is left to

*Sensors and Control Systems in Arc Welding.* Edited by Hirokazu Nomura.
Published in 1994 by Chapman & Hall, London. ISBN 0 412 47490 5

use his judgement, based on personal experience, to adjust the welding conditions to match deviations from standard that occur in the actual groove geometry. This is conducive to fluctuations in the resulting weldment quality, which thus depends heavily on the welder's skill and experience.

In order to minimize this dependence, the authors have developed a system whereby variations in groove geometry are automatically sensed, and the resulting signals are translated by microcomputer into compensating commands to adjust the welding conditions, and thus reduce the need for human intervention and personal judgement in welding operations.

## 22.2   AUTOMATIC CONTROL OF WELDING CONDITIONS

It has been proved that the deviations in root face thickness and groove angle within the ranges currently encountered in practice do not require modification of welding conditions [1]. However, root gaps in the groove, when welding conditions have been specified for zero gap, tend to result in excessive reverse bead and insufficient surface reinforcement. In practice, the operator is required to adjust the welding current and speed to match the root gap, which he must observe constantly during welding.

### 22.2.1   Relationship of welding conditions to reverse bead formation

Inagaki *et al.* [2] have derived the correlation between groove configuration and penetration from heat conduction theory, to define a quantity which they termed 'joint capacity' (JC). For one-side welding, as shown in Figure 22.1, JC is represented by the cross-sectional area required to be melted in order to produce a good reverse bead.

The reverse bead width is governed largely by the dynamic condition of heat generated by the arc, and it has been shown [2], [3] that a linear relationship can be obtained between the value of JC and the welding current to produce a reverse bead of given width. Applying this theory to three-electrode welding with flux copper backing, establishment of the quantitative relationship between root gap width and JC for obtaining a constant reverse bead width, and of the concomitant relationship between JC and the current supplied to the first electrode, should provide a means of controlling the welding conditions as a direct function of root gap.

The plots in Figure 22.1 show clearly how an increase of gap width $G$ is related linearly to a decrease of JC, in keeping with which a linear reduction of $I_1$ effectively maintains the reverse bead width well within acceptable limits. Moreover, the macrostructures reveal that the weld metal structure has been properly modified with the reduction brought to $I_2$ in keeping with $I_1$, where $I_2$ and $I_1$ are the currents on the second and first electrodes respectively.

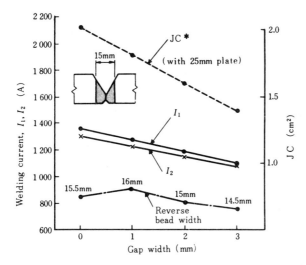

**Fig. 22.1** Relationship between each root gap and welding current: $I_1$ and $I_2$, currents on first and second electrodes; JC, joint capacity calculated for 15 mm reverse bead width.

### 22.2.2 Relationship of welding conditions to face bead formation

For a weld groove of given form, angles and root face thickness, the groove cross section $S_g$ is related to plate thickness and root gap. This should serve in controlling the total amount of weld metal that must be deposited by the three electrodes, and which would be controlled by adjusting their welding current and speed.

From the relationship between input heat and rate of electrode melting given by Wells [4], the deposite cross section [1] is given by

$$S_d = \frac{3}{20} \cdot \frac{1}{pcT_M} \cdot \frac{0.24\eta(E_1I_1 + E_2I_2 + E_3I_3)}{v} \tag{22.1}$$

where $\rho$ = specific gravity (g/cm$^3$), $c$ = specific heat (cal/g°C), $T_M$ = melting temperature of weld metal (°C), $\eta$ = thermal efficiency of welding.

Of the welding conditions, the current from the first electrode, $I_1$, is determined by the joint capacity JC; that from the second electrode, $I_2$, must be matched by $I_1$, to ensure the same modification of the weld metal structure whatever the quantity of metal deposition by the first electrode. Thus, provided that the welding voltages $E_1$, $E_2$ and $E_3$ can be maintained at levels matched to their respective currents, the total heat input to be adjusted to match variations in the deposit cross-section $S_d$ could be controlled by adjusting the welding speed $v$ and the current $I_3$ from the third electrode. Of these two values, $v$ is known not to affect the reverse bead significantly, so long as it is held within reasonable limits [2]. The plots of $S_d$ in Figure 22.2, calculated using Equation (1) assuming $\eta = 0.9$, $c\rho = 1.5$ cal/(cm³ °C) and $T_M = 1500$°C, further indicate good parallelism, with $S_g$ maintaining a distance corresponding to the reinforcements on both sides, amounting to 2–3% of $S_g$.

Consequently, increasing $I_3$ and decreasing $v$ in a certain proportion to $G$, represented by a corresponding increase of $S_g$, can be considered to provide the necessary conditions for ensuring a good weldment.

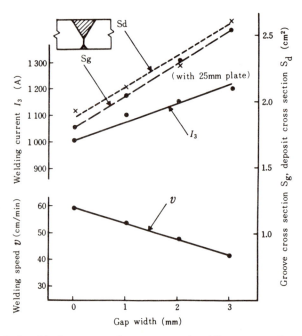

**Fig. 22.2** Relationship between each root gap and welding current $I_3$, welding speed and cross-sectional area.

### 22.2.3   Formula for determining the welding conditions

Based on the foregoing observations, formulae to govern $I_1$, $I_2$, $I_3$ and $v$ were established, which for 25 mm thick plate, for example, are as follows.

$$I_1 \,(\text{A}) = 1350\,\text{A} + K_1 \; G \;\text{mm} \; (K_1 = -85)$$

$$I_2 \,(\text{A}) = 1300\,\text{A} + K_2 \; G \;\text{mm} \; (K_2 = -75)$$

$$I_3 \,(\text{A}) = 1000\,\text{A} + K_3 \; G \;\text{mm} \; (K_3 = +70)$$

$$v \,(\text{cm/min}) = 60 \,\text{cm/min} + K_v \; G \;\text{mm} \; (K_v = -5.5)$$

(2)

where $K_1, K_2, K_3$ and $K_v$ are constants given individually for plates of different thicknesses.

## 22.3 SENSING SYSTEM

The sensor used to detect groove position and configuration required for the envisaged control system has to fulfil the following functions:

1. measuring the root gap $G$;
2. determining the position of the groove axis;
3. detecting the end point of the weld seam.

A laser diode combined with a transistorized linear position sensing detector was adopted.

The detected signals undergo preliminary treatment within the sensor unit at each scan, and are then transferred to the control unit. The control unit registers and processes the resulting data from the sensor unit, and derives the mean values of five consecutive scans.

## 22.4 CONTROL SYSTEM

### 22.4.1 System configuration

The control system is constituted as indicated in Figure 22.3. The requisite welding conditions are given by linear equations established separately for different plate thicknesses at intervals of 1 mm and filed in the control

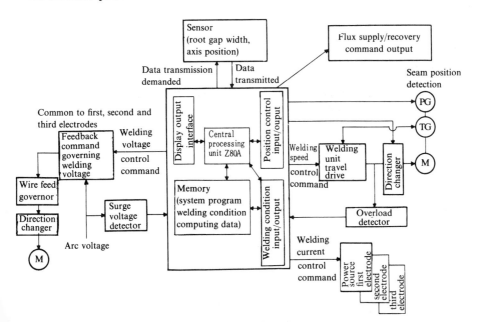

**Fig. 22.3** Block diagram of control system: M, motor; PG, pulse generator; TG, tachogenerator.

unit memory. The relevant plate thickness is communicated manually beforehand to preset the welding condition program.

Once the welding operation is started, the system can be left on its own until completion of the weld. The prototype machine is shown in Figure 22.4.

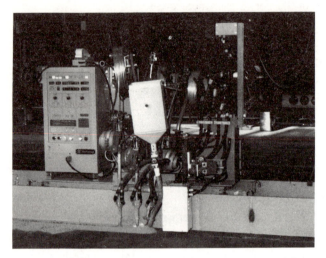

**Fig. 22.4** FCB (flux copper backing) one-sided submerged arc welding equipment.

### 22.4.2   Quality of weldments obtained

The weldments obtained in laboratory test and on actual plating for ship construction are presented respectively in Tables 22.1 and 22.2. The reverse bead width is in all cases stable within the range 14.5–16 mm.

## 22.5   CONCLUSION

Linear equations were established in consideration of the mechanism of bead formation. They relate to root gap width and the proper values of welding speed and of the welding currents to be supplied to the respective electrodes. These equations agreed well with the expression given by Wells [4].

The linear equations thus established are filed in the computer memory for different plate thicknesses in intervals of 1 mm, to be called out at the start of the welding operation to control the welding conditions.

A laser diode and beam detector system has been devised to detect the position of the groove axis and to discern the root gap width as well as other pertinent data on groove configuration, to provide the requisite information for automated welding.

**Table 22.1** Experimental welding results

| Plate thickness (mm) | Root gap (mm) | Reverse bead width (mm) | Macrostructure |
|---|---|---|---|
| 16 | 0 | 15.5 | |
| | 2 | 16.0 | |
| 25 | 0 | 14.5 | |
| | 2 | 15.0 | |
| 35 | 0 | — | |
| | 2 | — | |

The prototype machine has proved capable of the envisaged functions from welding start to termination. Upon application to the joining of plates for a ship, satisfactory results were obtained consistently with no remedial operation required on the weldments produced.

The incorporation of this machine in current shop practice should not fail to contribute appreciably to elimination of weldment quality fluctuations arising from differences in operator skill, and consequently to the stabilization and enhancement of weldment quality as well as to the saving of manual labour.

**Table 22.2** Welding results in actual works

| Plate thickness (mm) | Root gap (mm) | Reverse bead width (mm) | Bead shape (dimensions in mm) |
|---|---|---|---|
| 12.5 | 0 | 15 | 23 / 1.5 / 12.5 / 15 / 3 |
| 15.0 | 0.5 | 14 | 26 / 2 / 15 / 14 / 3 |
| 17.0 | 1.8 | 15 | 30 / 2 / 17 / 15 / 3 |
| 18.5 | 1.5 | — | 33 / 2 / 18.5 / 15 / 3 |
| 22.5 | 0 | 16 | 31 / 1 / 22.5 / 16 / 3 |

## REFERENCES

1. Kenko, S., Okui, M., Fujimoto, A., Iwabe, T., Sakiyama, K. and Mino, M. (1984) Automatic control system of one-side submerged arc welding by flux copper backing process. IIW document XII-827-84, 212-593-84.
2. Inagaki, M., Okada, A., Terai, K. and Okada, K. (1968) On the relation between groove configuration and penetration in submerged arc welding. *Trans. Res. Inst. for Iron, Steel and Other Metals*, **11** (5), pp. 27–41.
3. Okada, A. (1979) Selecting optimized conditions for one-side welding. Welding Institute of Japan Research Committee on Arc Physics.
4. Wells, A.A. (1952) Heat flow in welding. *Welding Journal*, **31**, pp. 263s–67s.

# CHAPTER 23

# Touch sensor and arc sensor for arc-welding robots

*Jun Nakajima, Takeshi Araya,*
*Shinnichi Sarugaku, Kooji Iguchi*
*and Yuuji Takabuchi*

The wire touch sensor and arc sensor have the advantages of compactness requiring no detectors around the welding torch, and of being unaffected by heat and arc light. Currently, these kinds of sensor are widely used with welding robots. However, they generally have some problems and limitations. To alleviate these problems, Hitchai have developed new control methods for wire touch sensors and arc sensors, and have implemented them on Hitachi welding robots.

## 23.1 INTRODUCTION

The wire touch sensor enables the position of a workpiece to be determined from the position of the robot axes when the welding wire contacts the work surface and current flow is detected. It is convenient and useful for detecting the position of the workpiece, because the wire itself, which does not need an extra sensing device around the welding torch, is used as a sensor. This sensor is applied for many kinds of sensing, such as detection of the welding start point for the arc sensor which utilizes welding arc phenomena or detection of groove faces and gaps.

However, the touch sensor introduces a problem of reduction of production efficiency, since detection of the position of a complicated workpiece requires many sensing actions prior to or during the welding operation.

Arc sensors also have positive features, but can be affected by disturbances due to welding arc phenomena.

*Sensors and Control Systems in Arc Welding.* Edited by Hirokazu Nomura.
Published in 1994 by Chapman & Hall, London. ISBN 0 412 47490 5

To improve these problems, a high-speed touch sensor has been developed, and satisfactory results obtained. We have also developed an arc sensor which filters out undesirable disturbances, frequency components, from the welding current.

This chapter describes the principle and performance of the wire touch sensor and arc sensor, and satisfactory practical applications.

## 23.2   PRINCIPLE AND PERFORMANCE OF WIRE TOUCH SENSOR

### 23.2.1   Principle and detecting operation

The wire touch sensor detects a welding start point and end point on a workpiece by utilizing the current between the wire and workpiece. Voltage is supplied between them and current is detected when the wire touches the workpiece. The value of the current is adjusted to the minimum value for sensing in order to ensure safety in operation.

Figure 23.1 shows the detection of the welding start point for a fillet joint. Points $P_1$–$P_4$ for the workpiece location A are taught in teaching mode.

Then, in playback mode, if the workpiece has moved from A to B, the robot moves along the taught path ($P_1$–$P_2$) until touching is detected by the sensor at point P. This is the sensor 'search' routine. The robot controller then calculates the shift value $S$ which is the distance between the actual workpiece point P and the taught point $P_2$. Instead of advancing to the taught point $P_4$, the robot advances to the actual welding start point $P'_4$ according to the calculation, $P_4 + S$.

Accordingly, it is necessary to improve the travelling speed during sensing in order to shorten the sensing time and to improve production efficiency. The travelling time has a significant influence upon workpiece detection time and production efficiency, especially when the workpiece position is

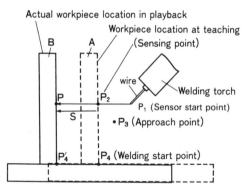

**Fig. 23.1** Detecting operation.

detected three-dimensionally or the workpiece has many welding lines detected by the sensor. Generally, the sensing speed $V_s$ is given by

$$V_s = G \, \Delta D$$

where $G$ is the loop gain (amplification rate) of the servo amplifier, and $\Delta D$ is the length of deviation (overrun distance after touch sensing). If the loop gain $G$ is set too high, stable control cannot be performed owing to generation of vibration. If the deviation $\Delta D$ is set too high, the detection accuracy decreases, since the wire is bent on contact. The sensing speed $V_s$ is less than 20 mm/s, when it is calculated with $\Delta D \leqslant 1$ mm and other control parameters. This value does not satisfy the requirement for good production efficiency. The wire touch sensor is therefore modified to shorten the detection time without wire deformation, even if the sensing speed is set high, by reversing the robot command direction as soon as the wire touches the workpiece.

Figure 23.2 shows the block diagram of the control algorithm. As soon as the wire touches the work, the contact treatment block outputs the maximum speed in the opposite direction to the direction of search. The current command to the motors become very high by this control. Since this output might damage the amplifiers or motors, the torque command to the motors is limited by the limiter, whose output is sent to the servo amplifiers. The contact treatment block also monitors the current speed signal which is input to the differential calculation block. If the current speed becomes zero or opposite to that of sensing movement, the robot stops outputting the maximum speed command.

Figure 23.3 shows the principle of the sensing movement obtained from the algorithm described above. Let the position of the workpiece be at B in Figure 23.1 which is denoted as $X_W$ in Figure 23.3. When the touch point P is detected at $t_0$, the speed command $S_v = -V_{max}$ is output from the contact treatment block as in Figure 23.3(b). The speed decreases sharply because of the negative speed command until it becomes 0 ($t_1$) as shown in Figure 23.3(b). The position command is ignored during this period (from $t_0$ to $t_1$). The position $X_W$ detected at $t_1$ is used as position command during the period.

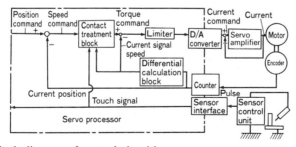

**Fig. 23.2** Block diagram of control algorithm.

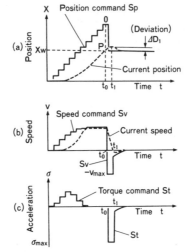

**Fig. 23.3** Principle of sensing movement.

As a result, it is possible to shorten the deviation $\Delta D$, and to apply a high search travelling speed (max. 50 mm/s). This shortens the time required to perform a search task.

### 23.2.2   Performance

Figure 23.1 shows the principle of detection of work piece movement in one dimension. A touch sensor is usually used to detect movement in three dimensions. It applies not only a parallel shift function to detect horizontal and vertical dislocation, but also a rotation shift function to shift the direction of the welding line.

**Table 23.1** Specifications of touch sensor

| Item | Specifications |
| --- | --- |
| Detection method | Wire touching |
| Surface of workpiece and wire | Resistance on surface of a material, less than 10 kΩ |
| Abnormality detection | Breaking of wire |
| Other functions | Warning monitor of touch sensing |
| Travelling speed (during sensing) | Approaching speed 1–50 mm/s<br>Leaving speed 1–1000 mm/s |
| Position shift by touch sense | Parallel shift (multidirections)<br>Three-dimensional rotation of line |
| Auxiliary functions | Wire extension adjustment function<br>Wire reverse feed function<br>Touch sense limitation (10–999 mm)<br>(Error detection accuracy, ±5 mm) |
| Number of shift registers | 99 |

The touch sensor and arc sensor are usually used together. The touch sensor detects a welding start point and the arc sensor follows a welding line.

Table 23.1 shows the specification of the developed touch sensor. The maximum travelling speed, 50 mm/s, is the main feature. The speed can be selected according to the characteristics of the workpiece. Automatic extension adjustment and wire negative feed are also installed (as auxiliary functions) in welding robots to improve detection accuracy by keeping the wire extension constant. Further, 99 shift registers, which memorize shift values detected by the touch sensor, are used to correct the welding torch position. They are useful for large and complicated workpieces.

## 23.3   PRINCIPLE AND PERFORMANCE OF ARC SENSOR

### 23.3.1   Principle of Arc sensor and system configuration

#### (a)   Principle

The general method of arc sensor used makes the welding torch weave, and detects the welding current in synchronization with the weaving. Further, it determines the position of the welding line according to the change in the welding current. Various methods have been considered for finding the shift of the torch position (weaving centre position) from the welding line [1].

#### (b)   Disturbance factors

The disturbance factors affecting an arc welding sensor include arc phenomena, fluctuation of the robotic mechanism during torch weaving (behaviour of the wire tip), and electrical noise. In particular, the arc phenomena constitute an important disturbance factor, because they cause different droplet transfer (such as short arc, globule, and spray) depending on the wire materials, welding power sources and welding conditions. Therefore, accurate positional information distinguished from disturbance must be appropriately obtained [2].

#### (c)   Arc sensor system

The control algorithm for the arc sensor described in this chapter filters the frequency components of the welding current. The algorithm aims at welding line profiling control with high following precision. This is accomplished by extracting particular frequency components from the variable welding current waveform and using them for correction of welding torch position [3].

The hardware system configuration is shown in Figure 23.4. The control unit of the arc sensor consists of a sensor unit and an automatic current control (ACC) section. In the sensor unit, high-frequency noises are rejected

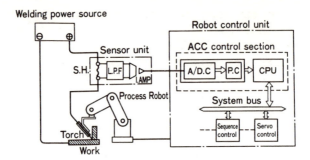

**Fig. 23.4** System hardware configuration: LPF, low-pass filter; I.AMP, insulated amplifier; ACC, automatic current control; ADC, analogue-to-digital converter; PC, photocoupler.

by a low-pass filter (LPF) from the welding current detected at a shunt. The current signal is then transmitted to the ACC section through an insulated amplifier (I.AMP). The ACC section converts the current signal into the corresponding digital signal by A/D conversion. It also makes electrical insulation for the signal using a photocoupler (PC), and executes an appropriate algorithm to output signals for correction to the torch position correction memory (which is shared with the robot control section). The robot control section reads the corrected signals to control the torch position.

The control algorithm extracts the frequency component with period equal to the weaving period. It also extracts the double frequency component with period equal to half the weaving period. This process calculates the control outputs using the frequency components for the correction of the torch position (or weaving centre position).

Welding current wave, current phases, and frequency power spectra corresponding to several shifts from the welding line are shown in Figure 23.5.

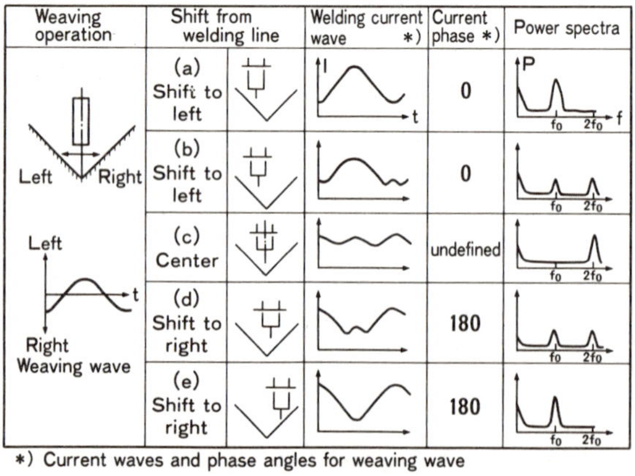

| Weaving operation | Shift from welding line | Welding current wave *) | Current phase *) | Power spectra |
|---|---|---|---|---|
| | (a) Shift to left | | 0 | |
| | (b) Shift to left | | 0 | |
| | (c) Center | | undefined | |
| | (d) Shift to right | | 180 | |
| | (e) Shift to right | | 180 | |

*) Current waves and phase angles for weaving wave

**Fig. 23.5** Power spectra of welding current according to shift from welding line. *Current waves and phase angles for weaving wave.

As shown by (c) in the figure, the frequency power spectrum of the welding current is high at the double weaving frequency component when the torch operates at the centre of the bevel joint.

If the weaving centre position is shifted from the welding line to the left as in (a) or (b), or to the right as in (d) or (e), the frequency power spectrum becomes high at the weaving frequency component. Accordingly, whether the weaving centre position is shifted from the welding line can be determined by the frequency components. In addition, the direction of the shift can be found by checking the phase of the current detected for the weaving wave as shown in Figure 23.5. The frequency components are detected by digital Fourier transformation (DFT).

The arc sensor control algorithm based on the principle described above is shown in Figure 23.6. In the figure, graph (1) shows the operation of the torch and graph (2) shows the corresponding welding current wave (filtered output). Graph (3) was obtained by DFT of the welding current values in graph (2).

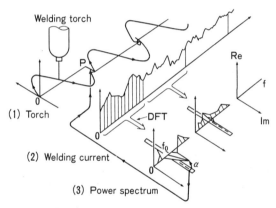

**Fig. 23.6** Control algorithm for arc sensor.

The processing overview will now be described. First, the welding current $I_n$ is collected repeatedly during a single weaving period of the torch. Then, at the end of the period, DFT is carried out for the sampled current $I_n$.

The values of the frequency power spectrum $P(f)$ can be described by the following equation:

$$P(f) = \int I(t) \cdot e^{-i\omega t}\, dt \approx \sum I_n \cdot e^{-i\Delta\omega t}\, \Delta t \qquad (23.1)$$

where: $\Delta t$ = sampling period (10 ms); $\omega$ = angular velocity (rad/s), $\omega = 2\pi f_0$, $\Delta\omega = \omega/\Delta t$; $f_0$ = weaving frequency (Hz); $N$ = Sampling times ($= 1/f_0 \Delta t$).

As previously described, the weaving frequency component $P(f_0)$ has the greatest correlation with the positional shift of the torch. As the positional shift deviation increases, the value of $P(f_0)$ becomes larger. If the direction of the shift is altered, then the phase of the welding current against the weaving is reversed as shown in Figure 23.5. The phase angle (shown by $\alpha$ in Figure 23.6) can be obtained from

$$\alpha = \arctan \frac{\text{Im}(P)}{\text{Re}(p)} \tag{23.2}$$

where: Im = imaginary part of power spectrum $P(f)$; Re = real part of power spectrum $P(f)$.

The direction in which the torch is moved for correction of torch position is determined by the value of $\alpha$ in Equation (23.2). The sequence described above is executed on the basis of the values indicated for weaving. The delay elements taken into consideration are the delay of robot response and the delay time of the filters.

### 23.3.2  Performance

#### (a)  Example of inspected data

Figure 23.7 shows the frequency power spectrum for the cases of shift and non-shift of the torch from the welding line. The results indicate that the control algorithm of the arc sensor realized through the frequency-conversion method is effective for tracking control of welding lines.

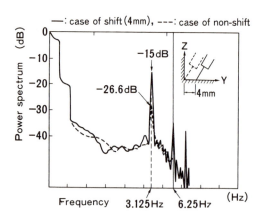

**Fig. 23.7** Frequency power spectrum.

#### (b)  Experimental results from a prototype system

Lap joint welding was performed with the arc sensor mounted on a robot (with six joint type axes), which resulted in a good tracking control. A photograph of the welding bead is shown in Figure 23.8.

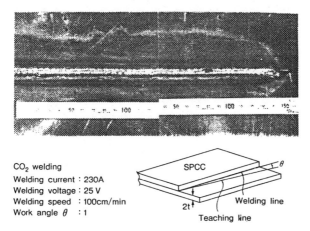

CO₂ welding
Welding current : 230A
Welding voltage : 25 V
Welding speed : 100cm/min
Work angle $\theta$ : 1

SPCC
Welding line
Teaching line
2t

**Fig. 23.8** Photograph of welding bead.

## (c) *Applied specification*

The range and specification applicable to a welding robot on which an arc sensor is mounted is shown in Table 23.2.

**Table 23.2** Range and specification applicable to arc sensor

| Condition | Description |
|---|---|
| Power source | 350 A |
| Current area | 120–350 A |
| Wire | 1.2 mm diameter solid wire |
| Shielded gas | MAG (Ar80% + $CO_2$), $CO_2$ |
| Weaving frequency | 2–3 Hz |
| Weaving amplitude | 2–5 mm |
| Torch angle | Foward angle 10–20° |
| Joint | Fillet, lap |
| Plate thickness | More than 2.0 mm |
| Gap | Less than 2 mm |

## 23.5 CONCLUSION

A wire touch sensor control method has been developed to shorten the detection time of workpiece mislocation or movement and to improve production efficiency. This method is lower in cost than the use of special brakes to stop motion when high-speed contact occurs.

An arc sensor control method which filters the frequency components of the welding current was developed. An experiment was performed with the sensor mounted on a welding robot. The results indicate that the method is extremely effective and provides high precision seam tracking.

## REFERENCES

1. Araya, T., Tsuji, M. and Araya, T. (1985) Advanced technology for arc welding robots. *Hitachi Review*, **34**, (1), 27–32.
2. Nomura, H., Sugitani, Y. and Suzuki, Y. (1981) Automatic seam tracking by arc sensor, in *Proc. National Meeting of the Japan Welding Society*, No. 28, May, pp. 184–5.
3. Nakajima, J., Araya, T., Tsuji, M. and Sarugaku, S. (1986) Arc sensor for welding line tracking applied to welding robot. IIW-XI-954-86.

# CHAPTER 24

# Application of arc sensors to robotic seam tracking

*Shunji Iwaki*

The arc sensor is one of the sensing systems which realizes the seam tracking function by using the physical properties of the welding arc. Compared with other seam tracking systems, arc sensors have the advantage that they do not require any special sensing device around the welding torch. This chapter introduces the principle of torch position control by arc sensor, and gives some examples of the welded result.

## 24.1 INTRODUCTION

In recent years, automation of welding operations, and especially their robotization, has spread rapidly, and applications have expanded to include not only steel sheet, particularly for motor vehicles, but also steel plate, particularly in construction machinery and steel articulated frames. More recently, robotization has spread to the welding of large structures as well, previously considered to be entirely within the domain of manual welding.

Generally, large structures have large errors caused by relatively low machining, assembly, positioning and other accuracies of their components, and are subject to relatively large thermal distortion during welding. Thus, if their welding is to be robotized, real-time tracking of the welding seam is essential. Arc sensing is one of the techniques by which this can be achieved. As the welding torch oscillates, the welding current and voltage vary. On the basis of these variations, the arc sensors control the torch position. For this reason, they carry out sensing with nothing else mounted on the welding torch, and hence assume an important position among the various sensors mounted on welding robots.

*Sensors and Control Systems in Arc Welding.* Edited by Hirokazu Nomura.
Published in 1994 by Chapman & Hall, London. ISBN 0 412 47490 5

This chapter describes the principle of the use of arc sensors to control torch position, and gives a few examples of their application.

## 24.4  PRINCIPLE AND CONSTRUCTION

Robotized welding of large structures is usually accomplished by the GMA welding process in which carbon dioxide gas, or a mixture of argon and carbon dioxide, is used. The welding power supply is of a constant voltage type and the wire feed speed is constant. Under these conditions, the welding current $I$ varies with the extension: the distance $l$ between the contact tip and the base metal. Figure 24.1 shows that, as $l$ increases, $I$ falls almost linearly. In the GMA welding process, in which this linear relationship is utilized, variations in welding current occurring as the torch oscillates over the welding seam are used as input signals to control the torch position.

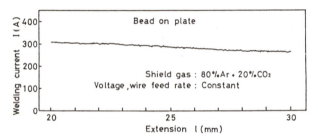

**Fig. 24.1** Relationship between welding current and extension.

By contrast, in the GTA welding process, the torch position is commonly controlled by using as input signals variations in arc voltage that are caused by variations in arc length with torch oscillation. This chapter concentrates on torch position control in GMA welding.

Figure 24.2 shows the principle of torch position control in flat fillet welding. Figure 24.2(a) illustrates torch oscillation and variation in the extension, while Figure 24.2(b) shows the corresponding variations in welding current. The broken line in (b) represents the case where there is no deviation between the welding seam and the torch's targeting position, while the solid line represents the case where there is some deviation. If there is no deviation, then we can divide each of the right- and left-hand halves of one cycle of oscillation into quarter-cycles, and let the mean current values for the two quarter cycles be $I_L^*$ and $I_R^*$. Then, since the curve is geometrically symmetric, the following equation holds:

$$I_L^* - I_R^* = 0 \tag{24.1}$$

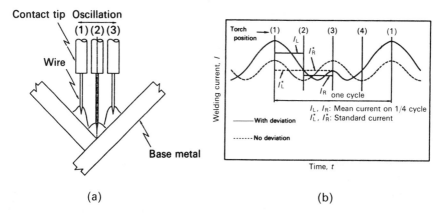

**Fig. 24.2** Principle of torch position control in flat fillet welding. (a) Torch oscillation. Torch position: 1, left edge; 2, centre; 3, right edge. (b) Current variation.

By contrast, if there is a deviation, then assuming the mean current values for the two quarter cycle parts to be $I_L$ and $I_R$, a deviation will appear, as shown below.

$$\Delta I = I_L = I_R \neq 0 \qquad (24.2)$$

Thus it is necessary to correct the centre of the torch oscillation so as to cancel this deviation $\Delta I$. Let the distance of correction be $H$, given by

$$H = a\Delta I \qquad [= a(I_L - I_R)] \qquad (24.3)$$

where $a$ is a constant.

If a constant distance between the base metal and the torch is to be maintained at the same time, it is necessary to control the mean current for each half cycle so that it will become equal to the reference value (the mean current in the case with no deviation). The amount of correction in the torch's axial direction will be given by

$$(V = b(I - I^*) \qquad (24.4)$$

where $I = I_L + I_R$, $I^* = I_L^* + I_R^*$, and $b =$ a constant.

In horizontal fillet welding, the targeting position is of the offset along the horizontal metal, as shown in Figure 24.3(a). In such a case, even if there is no deviation between the weld seam and the targeting position, the mean of the current values on the upper and the lower part of a quarter cycle will not equalize, and the deviation will be given by

$$\Delta I^* = I_U^* - I_L^* \neq 0 \qquad (24.5)$$

where $I_U^*$ and $I_L^*$ are the means of the current values on the upper and lower parts of a quarter cycle, respectively. The plot of the welding current in this case is represented by the broken line in Figure 24.3 (b).

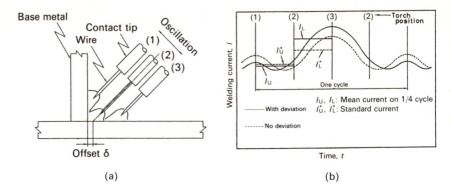

**Fig. 24.3** Principle of torch control in horizontal fillet welding. (a) Torch oscillation. Torch position: 1, upper edge; 2, centre; 3, lower edge. (b) Current variation.

By contrast, if there is a deviation between the welding seam and the targeting position, the plot of the welding current will be represented by the solid line in Figure 24.3 (b).

The deviation between the mean of the current values on the upper and the lower part of a quarter cycle is given by

$$\Delta I = I_U - I_L \neq \Delta I^* \tag{24.6}$$

Thus, in this case, correction must be accomplished in such a direction as will cancel the difference between $\Delta I$ and $\Delta I^*$. The amount of correction will be given by

$$H = c(\Delta I - \Delta I^*) \tag{24.7}$$

where $c =$ a constant. $\Delta I^*$ in this equation is the difference between the means of the current values on the upper and lower parts of a quarter cycle occurring if there is no deviation between the welding seam and the targeting position, as mentioned above. This difference is a function of the offset amount $\delta$, and it is difficult to predetermine its optimum value.

Thus it is necessary to determine the optimum value of $\Delta I^*$ by doing preliminary welding on testpieces or the like while changing the value of $\delta$ so as to achieve the required bead shape, and then storing the resulting optimum value of $\Delta I^*$, or by other appropriate methods. This method will give proper fillet welds of unequal leg length or other similar welds with ease. The amount of correction in the axial direction of the torch is given by Equation (24.4).

Figure 24.4 gives a diagram of the arc sensor system. The current signal passes through a low-pass filter intended for noise reduction, and is then converted into digital form at a given time interval. These digital values undergo integration to give the mean for each quarter cycle. These means are compared, and the amount of correction, $H$, is calculated using

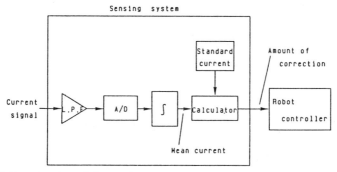

**Fig. 24.4** Construction of the sensing system.

equations (24.3), (24.4) and (24.7). The calculated value is transmitted to the robot controller for positional correction of the program.

## 24.3 EXAMPLES OF APPLICATIONS

Figure 24.5 shows two results of horizontal fillet welding; (a) for beads of equal leg lengths, and (b) for beads of unequal leg lengths. As can be seen from these photos, the arc sensor is also applicable where unequal leg lengths are required in fillet welding between plates of different thicknesses.

Figure 24.6 gives an example where the arc sensor was applied to the formation of a lap joint between sheets. The sheet was of SPCC material and 2.3 mm thick, and the welding speed was 20 mm/s. The frequency of

(a)                                      (b)

**Fig. 24.5** Typical bead profiles in fillet welding: (a) equal leg lengths; (b) unequal leg lengths.

**Fig. 24.6** Result on lapped joint.

torch oscillation was 5.5 Hz. The same control method as that applied to horizontal fillet welding was applied, with good results.

Finally, Figure 24.7 gives an example where the arc sensor was applied to welding a V-shaped groove while automatically adjusting the weaving width by sensing the walls of the groove. The testpiece had a groove with a root gap of 4 mm at its narrowest part and 15 mm at its widest part. The shielding gas was a mixture of Ar and $CO_2$. Simultaneous control of the weaving width and the welding speed provided defect-free, consistent beads. The control method employed differed slightly from that for fillet welding, but is exactly the same in principle, in that it uses variations in welding current with variations in the extension.

In addition to the above examples of applications, the arc sensor also has applications in the welding of single-sided bevel and J-shaped grooves, and vertical-up welding.

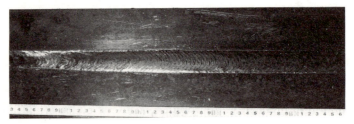

**Fig. 24.7** Application to a V-shaped groove.

## 24.4   CONCLUSION

This chapter has described the arc sensor's principle of control of the torch position, by comparing the means obtained through integration. This is a typical signal-processing method used with arc sensors. Other methods in use are diverse, including those that compare the instantaneous current values at the end point of torch oscillation; those that compare the frequency components of the welding current; those based on variations in the arc voltage at high resolution; and those which find the characteristics of the hysteresis loop formed by the torch position and the welding current by multivariate analysis. This indicates that existing arc sensors are not necessarily perfect, leaving room for improvement. If the performance of arc sensors is to be improved, signal processing techniques are important. Since arc sensors use a physical quantity associated with arcing, such as welding current or voltage, as input signal for control purposes, this quantity, and hence the welding arc, needs to maintain a stable condition. Thus improving the performance of welding equipment, including precision control of the welding current, improvement of the wire feed system and the power supply system for the welding wire, is also important.

When it comes to improving the quality of welds, there are some problems which cannot be coped with by an arc sensor alone. As mentioned above, an arc sensor works by using a physical quantity associated with the welding arc. It cannot detect the welding seam unless welding starts. Unless some method is used to detect the welding start position beforehand, there is a high possibility of welding defects occurring in the vicinity of that position. Various methods are available to detect the welding start position, but if one feature of an arc sensor (that there is no accessory required to be installed around the torch) is to be made use of, a wire touch sensor seems to be the optimum method. Quality robotized welding will only be possible when both sensors are combined.

Shin Meiwa Industry Co. Ltd. is doing research in many fields, including signal-processing techniques, welding equipment, and welding techniques, to improve the performance of arc sensors. The company has already developed touch sensors with a variety of functions. By improving the performance of both sensors in complete sensing systems, the company will contribute toward more efficient welding operations.

# CHAPTER 25

# Dynamic analysis of arc length and its application to arc sensing

*Etsuzou Murakami, Katsuya Kugai*
*and Hideyuki Yamamoto*

To apply arc sensors in high-speed welding, high-frequency weaving is required to maintain a good seam appearance, and to obtain suitable tracking capacity. In this study, the dynamic behaviour of the arc in weaving over 5 Hz is analysed in theory, and confirmed by experiment. An arc-sensing algorithm which can be applied from low- to high-frequency weaving is developed by using the result of that analysis. Furthermore, a digital filter for the arc current and an automatic parameter editing function are developed to make the algorithm suitable for practical use, and applied to an actual robot welding system.

## 25.1  INTRODUCTION

Tracking of seam deviations is indispensable to the promotion of automation in the arc welding process, and its realization requires sensors which can detect a seam at high speed and in real time. This has led to the development of sensors such as the laser sensor, and visual sensors such as the CCD camera. In comparison with these sensors, the arc sensor, which detects the position of a seam by measuring the arc characteristic physical values (current, voltage, etc.) when the torch is weaving, has the advantages of lower initial cost, easier setting of a welding position as no special device is needed around the torch, and no maintenance.

*Sensors and Control Systems in Arc Welding.* Edited by Hirokazu Nomura.
Published in 1994 by Chapman & Hall, London. ISBN 0 412 47490 5

However, because a higher weaving frequency is required for high-speed welding, the conventional arc sensor, which cannot provide a weaving frequency larger than 5 Hz, has the disadvantage that it cannot respond to high-speed welding over 1 m/min, thus limiting its applicability.

Therefore, analysing the arc welding phenomenon during high-speed weaving should be of interest in considering automation in arc welding.

## 25.2   ARC PHENOMENON IN WEAVING

When the welding torch is weaving in the groove, the arc current varies with the variation in the tip distance between the base and the metal. If there is no deviation between the seam and the tool centre point of the torch, the arc current shows symmetrical behaviour, as shown in Figure 25.1(a). If there is any deviation, it shows asymmetrical behaviour (Figure 25.1(b) and (c)). Thus, the arc sensor must detect the deviation in tool centre point relative to the seam from the variation in arc current behaviour.

When weaving is performed across the groove at a frequency of less than 5 Hz, as used for the conventional arc sensor, it is known that the relationship between weaving position $P$ and arc current $I$ shows a horizontal infinity current curve, as shown in Figure 25.1. An arc sensor utilizing the above phenomenon has been already developed.

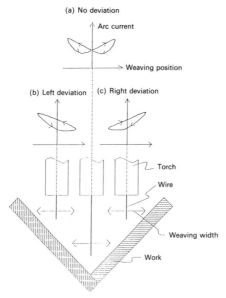

**Fig. 25.1** Arc current behaviour: (a) no deviation; (b) left deviation; (c) right deviation.

The phenomenon of a horizontal infinity curve suggests that the arc current does not merely vary with variation in the wire extension. It has been understood that, even at constant wire-feeding speed, the wire speed relative to the base metal side differs between when the tip of a welding wire is moved towards the edges of the joint groove and when it is moved towards the joint centre. The arc current tends to increase in the former case.

However, in an experiment to confirm the above phenomenon at 10 Hz weaving frequency, unlike the horizontal infinity curve shown in Figure 25.1, a phenomenon which reverses the path of a horizontal infinity curve shown in Figure 25.2 was observed. This shows that the basic theory is not sufficient to explain welding current behaviour during all weaving conditions of a horizontal infinity curve.

In order to review conventional arc sensor theory, a theoretical analysis was made to determine what mechanism causes the horizontal infinity curve and also what effects are caused by the characteristics of weaving frequency and of welding power supply, and the deviation in the targeted seam.

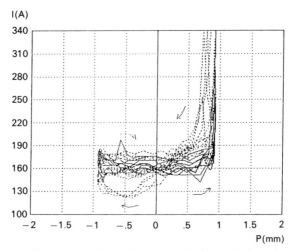

**Fig. 25.2** Measured arc current behaviour at 10 Hz (right deviation).

### 25.3   DYNAMIC BEHAVIOUR ANALYSIS

In general, the wire melting characteristic in arc welding is expressed by the following approximation.

$$V_m = aI_{av} + bl_x(I_{av})^2 \tag{25.1}$$

where $V_m$ = melting speed; $I_{av}$ = average arc current; $l_x$ = wire extension; and $a, b$ are constants.

The arc characteristics are given by

$$V_{av} = V_{ao} + \alpha I_{av} + \beta l_a \tag{25.2}$$

where $V_{av}$ = average arc voltage; $l_a$ = arc length; $V_{ao}$, $\alpha$ and $\beta$ are constants.
The characteristic of the welding power supply is given by

$$V_{av} = V_{eo} - RI^{av} - L\frac{dI_{av}}{dt} \tag{25.3}$$

where $V_{eo}$ = non-load voltage; $L$ = circuit impedance; $R$ = circuit resistance.
The following relationship is given between wire melt and wire feeding speed:

$$V_m = V_f - \frac{dl_x}{dt} \tag{25.4}$$

where $V_f$ = wire-feeding speed.
The tip to base metal distance $l_s$ is given by:

$$l_s = l_x + l_a \tag{25.5}$$

The relationship of average arc current $I_{av}$ to the tip to base metal distance $l_s$ is obtained as following, by using Equations (25.1)–(25.5).

$$K_2\frac{dI_{av}}{dt} + (K_1 + bK_2 I_{av}^2)\frac{dI_{av}}{dt} + \{a + b(l_s - K_0 + K_1 I_{av})I_{av}\}I_{av}$$

$$= V_f - \frac{dl_s}{dt} \tag{25.6}$$

$$K_0 = \frac{V_{eo} - V_{ao}}{\beta}, \qquad K_1 = \frac{R+a}{\beta}, \qquad K_2 - \frac{L}{\beta}$$

Now, on the assumption that a pool exists, the joint type is expressed by a quadratic curve as shown in Figure 25.3. The input terms on the right- hand side of Equation (25.6) at the weaved joint are obtained from the wire-feeding speed and weaving waveform to make the numerical analysis of Equation (25.6). As a result, the relationship between weaving position $P$ and average arc current $I_{av}$ was obtained as shown in Figure 25.4,

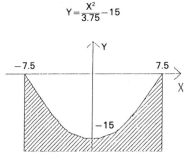

$$Y = \frac{X^2}{3.75} - 15$$

**Fig. 25.3** Model joint type: $y = x^2/3.75 - 15$.

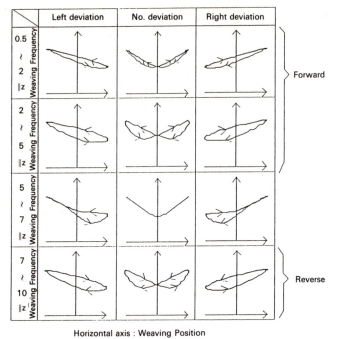

Horizontal axis : Weaving Position

Vertical axis : adjustment current

**Fig. 25.4** Weaving frequency and arc current behaviour: horizontal axis, weaving position; vertical axis, adjustment current.

simulating the phenomenon which reverses the path of a horizontal infinity curve at over 7 Hz of weaving frequency. It became apparent that Equation (25.6) could verify the phenomenon shown in Figure 25.2.

Further, in order to verify Equation (25.6), the numerical analysis was made for the same conditions as those used for experiments performed with the device shown in Figure 25.5. The result showed very good agreement, as shown in Figure 25.6 and confirmed that this theoretical analysis is valid.

The phenomenon which reverses the path of a horizontal infinity current curve at over 7 Hz as shown in Figure 25.4 can be explained by the secondary delay of average arc current $I_{av}$ increasing with an increase in variable frequency of the tip to base metal distance $l_3$, causing delay in the response of $I_{av}$.

It is known that the factor causing the delay of $I_{av}$ is $K_2$ in Equation (25.6), and circuit impedance $L$ in Equation (25.7). Therefore the numerical analysis was made, by taking $L$ as a parameter, on the weaving frequency which reverses the path of a horizontal infinity curve. As shown in Figure 25.7, the result showed that the frequency at which the path of a horizontal infinity curve reverses increases when circuit impedance $L$ increases.

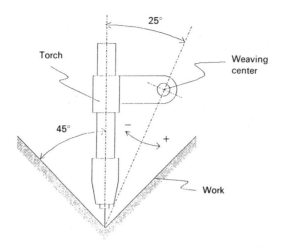

**Fig. 25.5** Experimental conditions for verifying the theory.

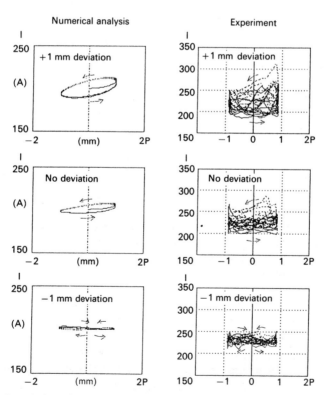

**Fig. 25.6** Comparison between numerical analysis and experiment.

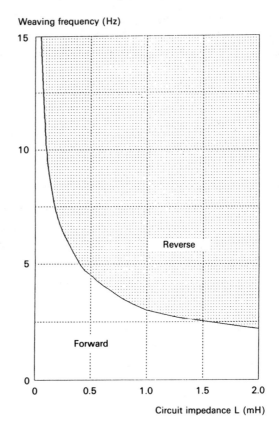

**Fig. 25.7** Circuit impedance.

## 25.4  APPLICATION TO ARC SENSORS

The above analytical results showed that consideration of a single aspect of arc current behaviour is not sufficient to respond to the wide range of weaving frequency and welding speed and, in particular, high-speed welding. The most direct method as an algorithm for an arc sensor which can cover a wide range of welding speed is to estimate the tip to base metal distance $l_s$ by calculating Equation (25.6). However, in practice, this method is not accurate since the noise element is amplified in the stage of calculating the secondary differentials of arc current. Therefore, another approach was considered for pattern-matching between the arc current behaviour obtained from Equation (25.6) and the measured arc current behaviour. However, since general pattern-matching is not practical owing to its large processing requirements, it was decided to employ an algorithm in which multiple parameters were introduced to express the characteristic of an infinity curve drawn by the arc current so that a linear combination of these patterns allows identification of variations in the path as patterns. The factors to make the linear combination of these parameters are determined based on Equation (25.6), and the input items includes weaving frequency, circuit impedance, welding condition and joint type. Among these items, because the joint type cannot be determined definitely due to effects of the melt pool, etc., it should be set by the multvariate statistical analysis technique so as to fit to the actual arc current. The algorithm described above is summarized in Figure 25.8.

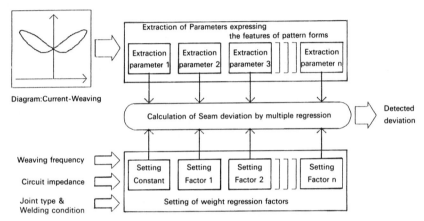

**Fig. 25.8** Theoretical expression-based algorithm for arc sensor.

## 25.5 DEVELOPMENT OF A PRACTICAL ARC SENSOR

### 25.5.1 Arc-current digital filter

The broad application of an arc sensor requires the achievement of consistent performance in different arc phenomena, whereas the theoretical analysis described above is basically applicable to a spray arc. Although the theoretical analysis has yet to be applied to a short arc, the arc current digital filter has been designed to provide the same sensing performance for a short arc as well as a spray arc, based on the assumption that Equation (25.6) is applicable to a short arc with a 'hypothetical averaged arc length'. Although the current for a short arc varies sharply, the waveforms can be measured by high-speed sampling at 10–50 kHz. Software operations were used to calculate a moving average of the waveforms in 10–20 ms, to convert them to 'false' arc current waveforms similar to those of a spray arc. It became evident that this method can provide nearly the same sensing performance for a short arc as for a spray arc.

### 25.5.2   Automatic parameter editing function

In general, since the sensing performance of an arc sensor is determined by the physical phenomena of the arc itself, the sensor needs to be adjusted to various factors such as welding conditions. However, because the adjustment work requires good knowledge of the functions of the sensor and of welding, it is difficult for the user to take up this work. In addition, in the algorithm shown in Figure 25.8, each factor may be modified due to a melting pool and gravity which are not accommodated in the theoretical expression. Therefore, in order to utilize the arc sensor fully, an automatic parameter editing function was developed to set these factors automatically.

First, in the condition where the welding system actually operates, teaching is effected without any deviation by using one actual workpiece as a testpiece. Test welding is then performed. The arc sensor samples the arc current behaviour, while the welding torch is performing a 'snaking' movement to the left and right of the centre of the taught seam, as shown in Figure 25.9. When sampling is completed, the arc sensor calculates the optimum value of each factor by multiple regression analysis in order to achieve agreement between the value when deviated from the seam and the detected value.

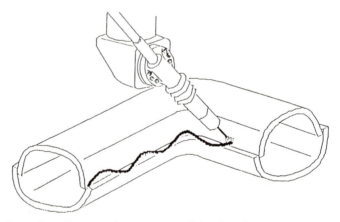

**Fig. 25.9** Operation of automatic parameter-editing function.

### 25.6   APPLICATION TO ROBOT WELDING SYSTEM

Many arc sensors have now been developed, but few of them are broadly applicable: in particular, in the field of thin plates. The reasons for the difficulty include: incapability in high-speed welding, numerous limitations in construction, low reliability in operation and difficulty in adjusting the sensor.

The 'Line Master' arc sensor unit for Daihen robots has been developed as an arc sensor which can be applied in practice by employing the arc sensor technology described earlier to solve the above problems. In addition, the Line Master embodies a touch sensor, which detects welding end points, to cover a variety of applications at the shop floors such as the correction of arc sensing start points. Table 1 shows the specifications for Line Master covering applications at welding speeds up to 1.5 m/min with 10 Hz weaving, which makes possible previously difficult seam tracking in downward welding.

**Table 25.1** Principal specifications for Line Master

| | |
|---|---|
| **Arc sensor** | |
| Maximum tracking speed | 15 cm/min |
| Trajectory accuracy | $\pm 1.0$ mm |
| Maximum correcting capability | 1 mm/s or 5° |
| Lead lag | $\pm 10\%$ |
| Torch angle | No limit (vertical downward possible) |
| Weaving | 1–10 MHz |
| Minimum plate thickness | T-joint: 1.2 mm / Lap fillet: 2.0 mm / V-joint: 4.5 mm |
| Function | Automatic parameter editing / Linear predicted tracking / Self-monitoring tracking |
| **Touch sensor** | |
| Searching | Stardard pattern search / Single-direction search / User's patterned search (option) |
| Search speed | 10–99 cm/min |
| Search accuracy | $\pm 1.0$ mm |
| Function | Calibration store/recall |

## 25.7 CONCLUSION

The new algorithm based on the dynamic analysis of arc phenomena has made it possible not only to realize an arc sensor applicable to high-speed welding but also to provide successful results in downward welding. In addition, the development of the peripheral technology has expanded the applicability of the arc sensor.

We are committed to the theoretical analysis for short arc as well as the analysis of the melting pool in the arc phenomenon in order to develop an advanced sensor of higher performance.

**Fig. 25.10** External view of Line Master.

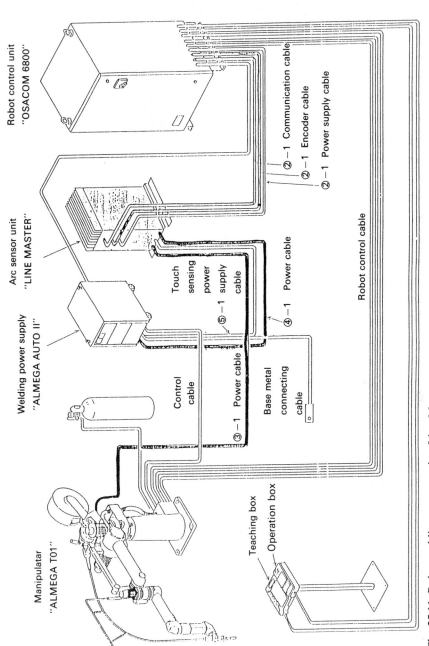

Robot control unit
"OSACOM 6800"

Arc sensor unit
"LINE MASTER"

Welding power supply
"ALMEGA AUTO II"

Manipulatar
"ALMEGA T01"

②—1 Communication cable

②—1 Encoder cable

②—1 Power supply cable

Touch
sensing
power
supply
cable

⑤—1
Power cable

④—1 Power cable

Robot control cable

Control
cable

Base metal
connecting
cable

③—1 Power cable

Teaching box

Operation box

**Fig. 25.11** Robot welding system using Line Master.

# CHAPTER 26

# Robot welding with arc sensing

*Hiroshi Fujimura, Eizo Ide and Hironori Inoue*

It is necessary for a teaching/playback arc welding robot to detect deviations of the weld line and correct the indicated locus; the so-called 'weld line sensing function'. The welding environment is exposed to high temperature, fumes, spatter, strong arc light, and electromagnetic noise. This causes difficulties and limitations for several types of sensor. However, the arc sensing method detects the weld line from variations in the arc, and has no additional external sensor. It is therefore an effective sensing method under severe environmental circumstances. The arc sensor described in this chapter calculates the tip–workpiece distance from the welding condition factors (welding current, welding voltage and wire feed rate) with a microprocessor and recognizes the amount of weld line displacement from the torch oscillation position. The practicality of this sensor has been confirmed by tests with an arc welding robot.

## 26.1   INTRODUCTION

Teaching/playback arc welding robots used for the medium- and large-scale structures encounter some problems. It is difficult to set the relative positions between the welding objects and the robot; it takes time for teaching and correction; and distortion during welding is large. There is therefore still a dependence on the workers' judgement, and robotization does not always realize the labour-saving and high efficience expected. To solve these problems, it is necessary for the robot itself to recognize the variation of the weld line and to correct the taught locus; the so-called 'weld line sensing function'. This chapter reports an example of an arc welding robot with an arc-sensing method which detects the weld line

*Sensors and Control Systems in Arc Welding.* Edited by Hirokazu Nomura.
Published in 1994 by Chapman & Hall, London. ISBN 0 412 47490 5

during welding using the arc welding characteristics and corrects the position of the torch on the basis of this information.

## 26.2 WORKING PRINCIPLE OF ARC SENSING

The consumable-electrode gas-shielded arc welding method used for current arc welding robots comprises the welding source, wire drive unit and welding torch. It generates the arc between the tip of the electrode wire, which is fed into the joint, and the workpiece. The wire and a part of the workpiece are fused under the shielding gas to form the weld bead. In this case the wire melts by arc heat and resistance heat. The arc heat is proportional to the welding current, and the resistance heat is a function of welding current and the wire extension. So there is a functional relationship between the wire melting rate, the welding current and the wire extension. Moreover, the arc length between the tip of the wire and the workpiece under a particular shielding gas, is determined by the welding current and the arc voltage. It is not possible to measure this arc voltage directly, so it is necessary to correct the voltage drop in the wire extension from the measurable contact tip–workpiece voltage (the welding voltage). The voltage drop and the resistance value vary with the wire extension, because the wire heats by passing current. From these welding pheomena we can obtain the contact tip–workpiece distance (the wire extension $L_E$ + the arc length $L_A$ by measuring the welding current $I$, the welding voltage $V$ and the wire feed rate $v$ (Figure 26.1).

Then, if the welding torch is oscillated across the joint groove, and the relationship of the oscillation position ($\theta$) and the contact tip–workpiece distance $L$ is determined, we can obtain the pattern of the groove shape, and judge the correct welding torch position both side to side and up and down (Figure 26.2).

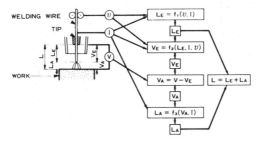

**Fig. 26.1** Flowchart for calculation of tip–workpiece distance: $v$, wire feed rate; $I$, welding current; $V$, voltage between tip and work; $V_A$, arc voltage; $V_E$, voltage drop through wire extension; $L_A$, arc length; $L_E$, wire extension length; $L$, tip–workpiece distance.

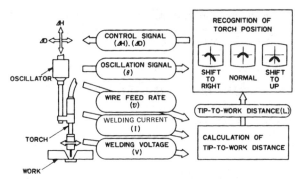

**Fig. 26.2** Operation principle of tracking system.

## 26.3   INTRODUCTION OF EQUATION

As described above, the arc sensor detects the arc line from the variation of the distance $L$ between the contact tip and the workpiece. As shown in Figure 26.3 $L$ is composed of the wire extension length $L_E$ and arc length $L_A$.

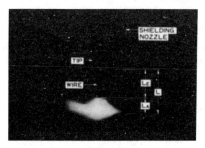

**Fig. 26.3** Arc state of gas–metal arc welding.

### 26.3.1   Equation for the wire extension length

The wire melting rate in consumable electrode arc welding is given experimentally by Halmoy as follows [1].

$$v_S = \frac{1}{H_0 - K_2 - \varepsilon}(\phi\delta + K_1 L_E \delta^2) \tag{26.1}$$

where $v$ is the wire melting rate; $\delta$ is the current density; $L_B$ is the wire extension length; $H_0$ is the heat contained in the weld metal (J/mm$^2$); $\phi$ is the equivalent voltage for melting; $K_1$, $K_2$ are wire constants; $\varepsilon$ is the calorie of wire at room temperature.

As regards Equation (26.1), Maruo and Hirata [2], Matsuda *et al.* [3] and Okada *et al.* [4] obtained the relative equation of wire melting rate $v_s$ in terms of welding current $I$, as follows:

$$v_s = K_3 I_a + K_4 I e^2 L_E \qquad (26.2)$$

where $I_a$ is the average current; $I_e$ is the effective current.

This wire melting rate, $v_s$, is a function of current, so in pulsed MAG welding this rate ought to vary with the pulse period. However, the time constant is large under the welding phenomenon, so the current seems to be smoothed in conditions where the pulse frequency is 20 Hz or those in which the arc is stabilized [3].

The pulse frequency of the common pulse MAG welding is above 60 Hz, so the arc achieves the equilibrium condition and the wire melting rate $v_s$ is regarded as the same as the wire feed rate $v$. So to change Equation (26.2), the wire extension $L_E$ is calculated with the average current $I_s$, the effective current $I_e$ and the wire feed rate $v$ from the following equation.

$$L_E = \frac{v - K_3 I_e}{K_4 L_a^2} \qquad (26.3)$$

### 26.3.2 Equation of the arc length

The relationship between the arc length $L_A$, the welding current $I$ and the arc voltage $V_A$ was formerly known as Ayrton's equation [5].

$$V_A = K_5 L_A + K_6 + (K_7 L_A + K_a) I \qquad (26.4)$$

This equation, which is obtained for the arc torch (the carbon electrode) experimentally, is used for the metal electrode [6]. So the arc length is calculated by the following equation.

$$L_A = \frac{V_A - K_6 - K_8 I_a}{K_5 + K_7 I_a} \qquad (26.5)$$

However, the arc voltage $V_A$ in Equation (26.5) cannot be measured directly, so it has to be estimated from the voltage drop at the wire extension, $V_E$, and the welding voltage $V$. $V_E$ is calculated as follows.

According to Ando *et al.* [5], the increase in temperature $\theta$ along the wire length $x$ after passing through the tip (electricity supplied point) is given by the following equation (Figure 26.4).

$$\theta = \frac{1}{\alpha} e^{at - 1} \qquad (26.6)$$

where $\alpha$ is the temperature coefficient difference between proper resistance and specific heat; $a = \alpha \delta^2 \eta_0 / (J \rho C_0)$; $C_0$ is the specific heat; $\eta_0$ is the specific resistance; $J$ is the mechanical equivalent of heat; $\rho$ is the density; and $\delta$ is the current density.

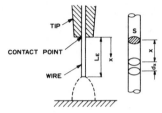

**Fig. 26.4** Geometry of wire extension.

Putting $t=x/v$ and $\delta=4I_e/\pi d^2$, where $d$ is the diameter of the wire, Equation (26.6) becomes

$$\theta=\frac{1}{\alpha}(e^{(K_9I_e^2x/v)}-1) \tag{26.7}$$

where $K_9=16\alpha\eta_0/J\pi^2J\rho C_0d^4$.

As shown in Figure 26.4 the resistance of wire of cross-sectional area $S$ and length $x$ is $r=\eta x/S$, so the resistance d$r$ of the minute length d$x$ is

$$dr=\eta_0(1+\beta\theta)dx/S \tag{26.8}$$

From Equations (26.7) and (26.8)

$$dr=\frac{4\eta_0}{\pi d^2}\left[\left(1-\frac{\beta}{\alpha}\right)+\frac{\beta}{\alpha}e\,K_gI_e^2x/v\right]dx$$

The resistance $R$ of the wire extension is calculated by integrating the above equation from $x=0$ to $x=L_E$:

$$R=\frac{4\eta_0}{\pi d^2}\left[\left(1-\frac{\beta}{\alpha}\right)L_E+\frac{\beta}{\alpha}\frac{v}{K_9I_e^2}(e^{K_9I_e^2L_E/v}-1)\right] \tag{26.9}$$

So the voltage drop at the wire extension, $V_E$, is calculated from

$$V=RI_a \tag{26.10}$$

As $V_E$ is calculated, as shown in Figure 26.1, the arc voltage $V_A$ is calculated from the measured welding voltage $V$ by the following equation:

$$V_A=V-V_E \tag{26.11}$$

So we can obtain the arc length $L_A$ by substituting the arc voltage $V_A$ and the average current $I_a$ into Equation (26.5).

### 26.3.3 Accuracy of the contact tip–workpiece distance calculation

In the previous section, we obtained the equation for calculating the contact tip – workpiece distance. We now evaluate the accuracy of this calculation by comparing it with the empirically measured value. The experiment was to record the average current $I_a$, the effective current $I_e$, the welding voltage $V$

and the wire feed rate for an arc generated between the fixed torch and the test plate traversed as shown in Figure 26.5. The contact tip–workpiece distance $L$ was kept constant.

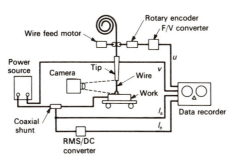

**Fig. 26.5** Measurement of wire extension $L_E$ and arc length $L_A$.

On the basis of these data and according to the flowchart in Figure 26.1, the relation between the contact tip–workpiece distance $L_c$, calculated from Equation (26.3), (26.5) and $L_c = L_E + L_A$, and the value of $L$ by measurement, $L_m$, was obtained.

The result is shown in Figure 26.6. Good correlation between $L_c$ and $L_m$ is seen with the average of errors $\bar{\chi} = -0.3$ mm and the standard deviation $\delta_{n-1} = 1.1$ mm. The error for the 25 mm contact tip–workpiece distance typically used for welding is within 5%, so it is judged that these calculation methods are able to be used.

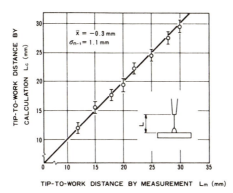

**Fig. 26.6** Calculation accuracy of tip–workpiece distance.

## 26.5  SENSING PERFORMANCE

The principle of the arc sensor is to judge the displacement at the welding arc in the joint groove from the relationship between the oscillation of the arc across the groove and the contact tip–workpiece distance ($\theta$–$L$ pattern) as described above. We observed the $\theta$–$L$ patterns obtained when different kinds of groove are welded, and obtained the relationship between this pattern and the amount of welding point displacement. As shown in Figure 26.7, the experiment was conducted for fillet or butt welding with the pulse MAG torch mounted on the oscillator. The welding condition factors ($I_a$, $I_e$, $V$, $v$) and the position of oscillation ($\theta$) were recorded. When the data were replayed at the calculation speed to the computer the $\theta$–$L$ pattern could be drawn.

Figure 26.8 shows the relationship between the torch path and the joint groove, giving varying amounts of welding point displacement $D$. The $\theta$–$L$ pattern obtained is shown for butt welding in Figure 26.9.

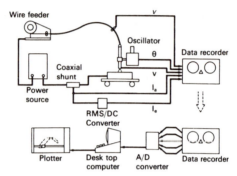

**Fig. 26.7** Method of analysis of $\theta$–$L$ pattern.

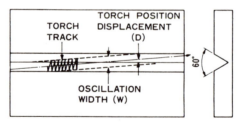

**Fig. 26.8** Dimensions of specimen for $\theta$–$L$ pattern analysis.

In Figure 26.9 the width of oscillation has values of 2 mm and 5 mm. When the width of oscillation narrows, the $\theta$–$L$ pattern does not show the groove shape correctly, but the trend in pattern correlates to the amount of welding point diplacement $D$, enabling it to be used for tracking of the joint groove.

### 26.4.1 Sensor control apparatus

On the basis of the results of the fundamental tests described above, we manufactured the sensor control apparatus.

The control apparatus uses a microprocessor and is composed of the oscillation position signal input section, the welding condition factor input

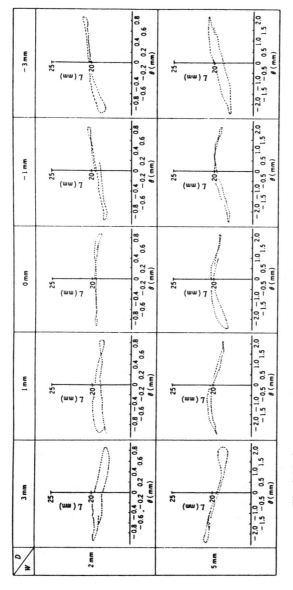

**Fig. 26.9** $\theta-L$ pattern for butt welding; $D$, offset of welding point; $W$, oscillation width.

section, the external set point input section, the error indication section and the correcting signal output section. Irrespective of oscillation amplitude, the oscillation position signal input section receives the position of the torch as a ratio from the oscillation or the robot control apparatus. As shown in Figure 26.7, the welding condition factor input section inputs the average current $I_a$, the effective current $I_e$, the welding voltage $V$ and wire feed rate $v$ which are necessary to calculate the contact tip–workpiece distance $L$. $I_a$ and $I_e$ are taken from a series coaxial shunt, the RMS value of $I_e$ is converted to d.c. and both are digitized by an A/D converter before passing to the computing section. $V$ requires A/D conversion and $v$ is the signal from an encoder fixed to the wire drive unit and fed directly to the computer. The oscillation position is taken at a trigger signal when the respective oscillation position $L$ is calculated. The external set point input section establishes the set point for control of the torch height and the shift quantity to obtain the correct bead shape and can obtain the required data from the robot control apparatus.

### 26.4.2   Application to the arc welding robot

The performance of the arc sensor was confirmed as discussed previously, then its application to the arc welding robot was tested. This robot has been developed for the welding of medium- and large-scale structures, and is shown in Figure 26.10. The arc sensor cannot detect the point of welding start prior to welding. However, this robot also has a wire touch sensor, so by using both sensors, perfect weld line tracking is possible. Sample welds are shown in Figure 26.11. Both butt and horizontal fillet joints have S-shaped seams with curvatures of about 300 mm radius, but good seam following and weld beads are achieved.

**Fig. 26.10** External view of arc welding robot.

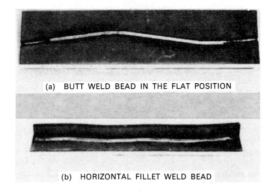

(a) BUTT WELD BEAD IN THE FLAT POSITION

(b) HORIZONTAL FILLET WELD BEAD

**Fig. 26.11** Weld bead produced by tracking controller.

## 26.5 CONCLUSION

For the purpose of improving the reliability of weld line sensing, which is an important factor for robotizing, we have developed a new type of sensor. The sensor developed in this research obtains the welding current, the welding voltage and the wire feed rate and by computation the absolute value of the contact tip–workpiece distance is determined as the basis for weld line search and control of torch height. Results of the application to the arc welding robot show high detection accuracy and achievement of reliable welding. In the future, on the basis of the method developed in this research, we will study possible further intelligence for arc welding robots.

## REFERENCES

1. Halmoy, E. (1979) Wire melting rate, droplet temperature and effective anode melting potential, in *Proceedings on Arc Physics and Weld Pool Behaviour*, p. 29.
2. Maruo, H. and Hirata, Y. (1980) Study on current controlled arc welding. Preprint of JIW 60th Technical Commission on Welding Processes (in Japanese).
3. Matsuda, F., Ushio, M. and Tanaka, Y. (1982) Particle transfer phenomena in pulsed welding. 90th Technical Commission on Welding Processes (in Japanese).
4. Okada, A., Yamamoto, H., Harada, S. and Nishikura, K. (1982) Pulsed MAG arc welding method by transistor power source. 86th Technical Commission on Welding Processes (in Japanese).
5. Ando, K. and Hasegawa, M. (1962) Welding arc phenomenon, Sanpoh publication (in Japanese).
6. Oshima, K., Abe, M. and Kubota, T. (1982) Sample data control of arc length in MIG pulsed arc welding. *Journal of the Japan Welding Society*, **51**(2), pp. 91–7.

# CHAPTER 27

# Arc sensing using fuzzy control

*Hiroshi Fujimura, Eizo Ide and Shuta Murakami*

One of the most important techniques in weld automation is weld-line detection and seam following. Compared with other weld-line sensing methods, such as visual sensors, the arc sensor does not need the attachment of an additional detector around the torch; weld-line detection is immediate; and the system is highly flexible. It is widely used to monitor automatic arc welding, including robot welding. However, the arc sensor uses the welding arc phenomena, and the parameters of the phenomena (including welding current, welding voltage and wire feed rate) are subject to external disturbances. The way these parameters are treated therefore influences the accuracy and reliability of the weld line detection. This chapter gives an example of the application of fuzzy control, which is said to be the emulation of human control which has considerable potential for many weld-line sensing applications.

## 27.1  INTRODUCTION

The most important techniques in weld automation are weld line recognition and following. Among these techniques the arc sensor, which does not need a detector attached to the torch, provides virtually immediate following, and the control system softness is used widely for automating the arc welding process. However, the arc sensor uses the arc phenomena and the parameters of the phenomena are affected by many external disturbances which influence the accuracy and reliability of weld line detection. In this chapter we apply fuzzy control, which is said to be the emulation of human control, for weld line sensing.

*Sensors and Control Systems in Arc Welding*. Edited by Hirokazu Nomura.
Published in 1994 by Chapman & Hall, London. ISBN 0 412 47490 5

## 27.2  WELD LINE SENSING BY FUZZY CONTROL

Fuzzy control has the characteristic that however difficult is the numerical model of the objective system, fuzzy information from operator experience, and the nature of the system, can model soft and adaptable control methods as the language and produce regulatory control algorithms of the IF–THEN form.

### 27.2.1  Working principle of the arc sensor

In this method [1] the arc is oscillated across the joint groove, and the contact tip–workpiece distance $L$ during the oscillation is determined, enabling the groove sectional shape and the correct torch position to be established. The distance between the contact tip and the tip of the wire is the wire extension $L_E$, the distance between the tip of the wire and the welding object is the arc length $L_A$, and the sum of $L_E$ and $L_A$ is the contact tip–workpiece distance $L$. The process of detecting the welding current $I$, the arc voltage $V$ and the wire feed rate $v$, which are the welding parameters, and calculating $L$ is shown in Figure 27.1, and the principle of weld line sensing control is shown in Figure 27.2. If the relationshp between the torch position information $\theta$ and $L$ is drawn in two-dimensional coordinates, the characteristic $\theta$–$L$ pattern for the groove sectional shape is obtained.

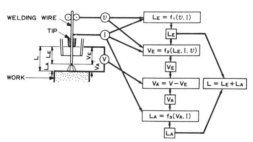

**Fig. 27.1** Flowchart for calculation of tip–workpiece distance: $v$, wire feed rate; $I$, welding current; $V$, voltage between tip and work; $V_A$, arc voltage; $V_E$, voltage drop through wire extension; $L_A$, arc length; $L_E$, wire extension length; $L$, tip–workpiece distance.

When the welding point displaces in the groove, this $\theta$–$L$ pattern changes, and the amount of welding point displacement is recognized from the asymmetry between the right and left halves of the oscillation. If $L$ is averaged in each $\theta$–$L$ pattern period, torch height control is possible.

### 27.2.2  Composition of the control system

The configuration of the control for weld line following is shown in Figure 27.3. This control system comprises the vertical control system, which

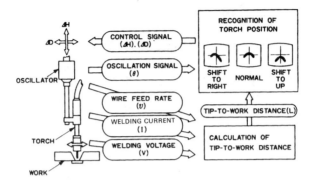

**Fig. 27.2** Operation principle of tracking system.

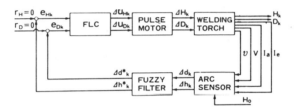

**Fig. 27.3** Block diagram of weld-line tracking control system: $k$, time (s); $r_H$, $r_D$, set points (mm); $e_{Hk}$, $e_{Dk}$, error (mm); $\Delta U_{Hk}$, $\Delta U_{Dk}$, manipulated variable; $\Delta H_k$, $\Delta D_k$, correcting quantity for welding torch position (mm); $H_k$, $D_k$, welding torch position after movement (mm); $v$, wire feed rate (mm/s); $V$, arc voltage (V); $I_a$, mean current (A); $I_e$, effective current (A); $\theta$, oscillation position signal; $\Delta h_k$, $\Delta d_k$, output of arc sensor (mm); $\Delta k_k^*$, $\Delta d_k^*$, controlled variable (mm).

corrects the up-and-down direction of the welding torch, and the horizontal control system which corrects the side-to-side direction: a two input – two output system. The line from $r_H$ to $\Delta h_k^*$ is the vertical control system and the line from $r_0$ to $\Delta d_k^*$ is the horizontal control system. The vertical and horizontal control systems have an individually designed FLC (Fuzzy Logic Controller). $r_H = 0$ is the target value which makes the deviation between the average of $L$ in each period of welding torch oscillation and the torch height set point ($H_0$) into zero. $r_D = 0$ is the target value which makes the horizontal deviation from the centre of the weld line into zero. To obtain the control amounts in the sample time ($\Delta h_k^*$ and $\Delta d_k^*$ from the target values), we have the control deviation, $I_{HR}$ and $I_{DR}$. If they are passed to the FLC, each operation amount, $\Delta \mu_{HK}$ and $\Delta \mu_{DK}$, is obtained. The outputs $\Delta h_K$ and $\Delta d_K$ are obtained from the arc sensor on the basis of the equation given below. Moreover, $\Delta h_K$ and $\Delta d_K$ are filtered through the fuzzy filter, the control amounts $\Delta h_k^*$ and $\Delta d_k^*$ are obtained, and they act as the feedback. The method by which the sensor outputs $\Delta h$ and $\Delta d$ are obtained during oscillation of the welding torch is shown. To obtain the $\theta$–$L$ pattern in Figure 27.4, the sampling of the

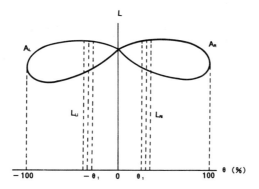

**Fig. 27.4** $\theta$–$L$ pattern.

contact tip–workpiece distance $L$ is performed between $+\theta\%$ and $-\theta\%$ of the oscillation position.

Then if the sample value of the right side of the $\theta$–$L$ pattern is $L_{Rt}$ and the sample value of the left is $L_{Lt}$, the following equation is obtained

$$A_R = \frac{t=\sum_1^m L_{Rt}}{m}, \qquad A_L = \frac{t=\sum_1^n L_{Lt}}{n} \qquad (27.1)$$

where $m$ is the sample number of the right side of $L$; $n$ is the sample number of the left side of $L$; $A_{Rt}$ and $A_L$ are the average values of $L$ in a cycle.

Next, the displacement values of the vertical and horizontal directions in a cycle, $\Delta h$ and $\Delta d$, are obtained by the following equations.

$$\Delta h = H_0 - \frac{A_R + A_L}{2} \qquad (27.2)$$

$$\Delta d = A_R - A_L \qquad (27.3)$$

where $\Delta h$ is the vertical displacement from the set point; $\Delta d$ is the horizontal displacement.

### 27.2.3 Fuzzy filter

The output of the arc sensor is based on the contact tip–workpiece distance which is obtained from the welding current and the welding voltage, etc. It is regarded as a fuzzy amount disturbed by noise. The composition method of fuzzy filtering by which the arc sensor output, including the noise, $\Delta h$ and $\Delta d$ are filtered is as follows. In order to obtain $\Delta h_K$ filtered at the sample time $K$, $\Delta h_{K-1}^*$, which is the value before $\Delta h_k^*$, and $\Delta h^* h_k^*$ are regarded as individual fuzzy numbers and the membership function is obtained.

The method by which $\Delta h_K^*$ is obtained from the two membership functions is shown in Figure 27.5. The $\Delta h_K$ membership function is more gentle than $\Delta h_{K-1}^*$ in this figure, which means that $\Delta h_K$ has greater fuzziness than $\Delta h_{K-1}^*$. More specifically, it shows that the reliability of $\Delta h_{K-1}^*$, considering all past data, gets higher evaluation than the current arc sensor output $\Delta h_K$. So the larger $x$ coordinate of the two intersections of the membership functions is selected as $\Delta h_k^*$.

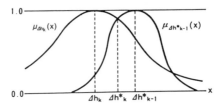

**Fig. 27.5** Procedure for obtaining $\Delta h_k^*$.

### 27.2.4   Design of FLC

On the basis of the configuration of the control system described in 27.2.1, we now describe a two input–output FLC design method. When the control amount is $\Delta h_K^*$, $\Delta d_K^*$, the operating amount is $\Delta \mu_{HK}^*$, $\Delta \mu_{DK}^*$, and the target value is $r_H$, $r_D$ at the same time $K$. The FLC is designed by using the following equations as FLC input variable.

$$\text{Vertical; } l_{HK} = r_H - \Delta h_K^* \text{ (control deviation),}$$
$$\Delta e_{HK} = l_{HK} - l_{H,K-1} \text{ (control deviation change)}$$
$$\text{Horizontal; } l_{DK} = r_D - \Delta d_K^* \text{ (control deviation),}$$
$$\Delta e_{DK} = l_{DK} - l_{D,K-1} \text{ (control deviation change)}$$

In this control system, the language control regulation (LCR) is composed as follows.

$$\text{LCR1: IF } e_{HK} \text{ is } P_H \text{ THEN } \Delta \mu_{HK} \text{ is } N_{H1},$$
$$\text{LCR2: IF } e_{HK} \text{ is } N_H \text{ THEN } \Delta \mu_{HK} \text{ is } P_{H1},$$
$$\text{LCR3: IF } \Delta l_{HK} \text{ is } P_{CH} \text{ THEN } \Delta \mu_{HK} \text{ is } N_{H2},$$
$$\text{LCR4: IF } \Delta l_{HK} \text{ is } N_{CH} \text{ THEN } \Delta \mu_{HK} \text{ is } N_{H2},$$
$$\text{LCR5: IF } l_{HK} \text{ is } P_D \text{ THEN } \Delta \mu_{DK} \text{ is } N_{D1},$$
$$\text{LCR6: IF } l_{DK} \text{ is } N_D \text{ THEN } \Delta \mu_{DK} \text{ is } P_{D1},$$
$$\text{LCR7: IF } \Delta l_{DK} \text{ is } P_{CD} \text{ THEN } \Delta \mu_{DK} \text{ is } N_{D2},$$
$$\text{LCR8: IF } \Delta l_{DK} \text{ is } N_{CD} \text{ THEN } \Delta \mu_{DK} \text{ is } P_{D2},$$

where P is positive, N is negative, subscribt H is vertical, and subscript D is horizontal.

LCR1 means that if equation $l_{HK} = -\Delta h_k^*$ is positive, i.e. the average height of the contact tip–workpiece distance at sample time $K$ is above the target value $H_0$, then the vertical operation amount $\Delta\mu_{HK}$ is negative, and the welding torch descends. LCR3 means that if the average height of the torch is far from the target value or if it approaches the target value from below, the torch descends. LCR5 means that if the equation $l_{DK} = -A_R + A_L$ is positive, then the torch shifts to the right of the centre of the weld line, and if the horizontal operation amount $\Delta\mu_{DK}$ is negative, then the torch is moved to the left. LCR7 means that if the torch deviation to the right side enlarges or the deviation towards the left decreases, then the torch is moved towards the left. Then if $l_{HK}$, $\Delta l_{HK}$, $l_{DK}$ and $\Delta l_{DK}$ are given as the input data at sample time $K$, the fuzzy reasoning harmonizes each regularity of LCR1–LCR4 and LCR5–LCR8. As a result, the vertical operation amount $\Delta\mu_{HK}$ and the horizontal operation amount $\Delta\mu_{DK}$ are obtained [2].

## 27.3 APPLICATION RESULTS USING FUZZY CONTROLLER

In order to compare the FLC described above, and the effectiveness of the fuzzy filter, the FLC without the filter and PI controller, a simulation and a welding test were performed.

### 27.3.1 Simulation test

The test conditions and workpiece shape in this simulation were as follows. The contact tip–workpiece distance $L$ was obtained by measuring the welding current, the welding voltage and the wire feed rate. It was calculated geometrically and contained normal noise (average value 0 mm, standard deviation 1.2 mm). Also, the sampling interval of $L$ was 8 ms for the actual arc sensor, but the sample numbers $m$ and $n$ were determined to approach this possible value in this test.

*(a)   Welding torch action conditions*

Oscillation amplitude: 30 mm
Oscillation number: 40 times/min
Welding rate: 1000 mm/min

*(b)   Setting of workpiece*

The L shape of the weld line was established as the workpiece shape, as shown in Figure 27.6.

### 27.3.2 Simulation test results

Four simulation tests were performed on test pieces as shown in Figure 27.6:

(a) FLC system with fuzzy filter;

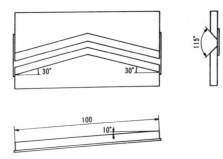

**Fig. 27.6** Testpiece for weld-line tracking simulation test.

(b) FLC system without filter;
(c) proportion integral (PI) controller system with exponential smoothing filter;
(d) PI controller system without filter.

The results are shown in Table 27.1. The evaluation of the results is based on the average value of deviation from the centre of the weld line, the standard deviation, the greatest and least value of the deviation (range) and the deviation in the curved line (curved point deviation). The results of further tests for the FLC system with the fuzzy filter (a) with conditions that the horizontal angles of the workpiece are 30° and 15° and the vertical angles are 10°, 0° and −10° are shown in Table 27.2.
    The following is obtained from Table 27.1.

1. The FLC system with the fuzzy filter performed best.
2. Comparing FLC with the PI controller, there is no significant difference in the control results for the vertical directions, but FLC is better for the horizontal directions.

**Table 27.1** Simulation test results on four control systems

|  | Average | Standard deviation | Range | Curved point deviation |
|---|---|---|---|---|
|  | (mm) | (mm) |  | (mm) |
| (a) Horizontal | 0.0027 | 0.1575 | −0.38–0.76 | 0.76 |
| Vertical | −0.0001 | 0.1200 | −0.37–0.32 | — |
| (b) Horizontal | 0.1865 | 0.4389 | −2.42–1.59 | 1.59 |
| Vertical | −0.1856 | 0.1664 | −0.89–0.26 | — |
| (c) Horizontal | −0.0007 | 0.3066 | −0.76–1.06 | 1.06 |
| Vertical | 0.0004 | 0.1225 | −0.37–0.33 | — |
| (d) Horizontal | −0.1440 | 0.6460 | −1.98–1.32 | 1.21 |
| Vertical | −0.1549 | 0.1957 | −0.94–0.38 | — |

(a) FLC system with fuzzy filter
(b) FLC system without filter
(c) PI control system with exponential smoothing filter
(d) PI control system without filter

**Table 27.2** Results of tests on FLC system with fuzzy filter

| Angle | | Average (mm) | Standard deviation (mm) | Range | Curved point deviation (mm) |
|---|---|---|---|---|---|
| Horizontal: | 30° | 0.0027 | 0.1575 | −0.38–0.76 | 0.76 |
| Vertical: | 10° | −0.0001 | 0.1200 | −0.37–0.32 | — |
| Horizontal: | 30° | −0.0044 | 0.1606 | −0.37–0.73 | 0.73 |
| Vertical: | 0° | −0.0058 | 0.0663 | −0.13–0.41 | — |
| Horizontal: | 30° | −0.0042 | 0.1581 | −0.38–0.75 | 0.75 |
| Vertical: | −10° | 0.0002 | 0.1208 | −0.38–0.49 | — |
| Horizontal: | 15° | 0.0018 | 0.1819 | −0.48–0.70 | 0.35 |
| Vertical: | 10° | 0.0002 | 0.1204 | −0.37–0.31 | — |
| Horizontal | 15° | −0.0155 | 0.2138 | −1.19–0.67 | 0.31 |
| Vertical: | 0° | 0.0032 | 0.0849 | −0.29–0.41 | — |
| Horizontal: | 15° | −0.0051 | 0.1868 | −0.48–0.69 | 0.31 |
| Vertical: | −10° | 0.0000 | 0.1158 | −0.38–0.39 | — |

3. Comparing the results of the control system with fuzzy filter to the results without, there is a large difference, so the effectiveness of the fuzzy filter is confirmed.

Moreover, from Table 2:

4. The FLC system with fuzzy filter obtains approximately constant results in spite of the changes of horizontal and vertical angle of the workpiece.

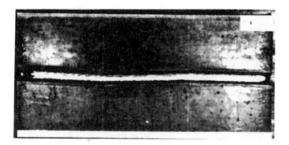

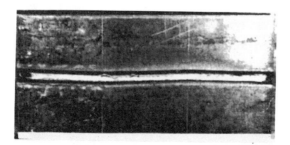

**Fig. 27.7** Appearance of weld bead: (a) PI control; (b) fuzzy control.

### 27.3.3   Arc welding test

Test welding using both PI control and fuzzy control was carried out. The PI control method has the moving average filter in the controller input, and the fuzzy control system has the fuzzy filter in FLC. The workpiece is set with the horizontal angle about 3°, the vertical angle about 2° and the weld line has two bends. The appearance of the bead obtained by PI control (a) and fuzzy control (b) are shown in Figure 27.7. Compared with fuzzy control, when using PI control:

1. the bead tends to 'snake' and the bead end is irregular;
2. the weld bead tends not to fill the groove completely.

As a result, it is confirmed that fuzzy control is better than PI control.

## 27.4   RESULTS

The efficiency of fuzzy control was confirmed by introducing it to the weld line sensing, and by the results of the simulation and the welding tests.

Fuzzy control was first applied to steam engine control by Mandani. Recently application has rapidly expanded to include automobile speed control, subway autodrive system and many others. This research has demonstrated the potential for application of fuzzy control to the welding process.

## REFERENCES

1. Fujimura, H., Ide, E. *et al.* (1983), Development of weld line tracking sensor for arc welding robot. *Mitsubishi Jūko Gihō*, **20**(6) 17.
2. Murakami, S. (1988) Arc welding fuzzy robot. *Journal of the Robotics Society of Japan*, **6**(6), 74–80.

# CHAPTER 28

# Development and application of arc sensor control with a high-speed rotating-arc process

*Hirokazu Nomura, Yuji Sugitani, Yukio Kobayashi and Masatoshi Murayama*

Because of problems such as the shortage of skilled welders, there is considerable demand for automation and robotization of arc welding processes. The high-speed rotating arc welding process easily achieves high-speed weaving (arc rotation) of 100 Hz. This process remarkably improves the sensitivity and responsiveness of the arc sensor, and has advantages for high-speed welding from its physical characteristics. This chapter describes the principle, features, and practical application of the process.

## 28.1 INTRODUCTION

Along with the rapid growth of production, various problems, such as the shortage of skilled welders, have grown recently in Japan. Therefore there is considerable demand for automation and robotization of arc welding processes. The high-speed rotating arc welding process developed by NKK has a number of advantages, including precise and responsive seam tracking by arc sensor control and high-speed welding ability due to its physical characteristics. High-efficiency, quality welding can be achieved.

This process has already been widely put into practical use, from narrow-gap welding of heavy thick plate to high-speed welding of thin plate. This chapter describes the principle, features and applications of the high-speed rotating arc process.

*Sensors and Control Systems in Arc Welding.* Edited by Hirokazu Nomura.
Published in 1994 by Chapman & Hall, London. ISBN 0 412 47490 5

## 28.2  OUTLINE OF THE HIGH-SPEED ROTATING ARC WELDING PROCESS

### 28.2.1  Principle

The principle of the process is shown in Figure 28.1. The electrode is fitted to the gear at a specific eccentricity via a bearing so as to allow rotation with its upper end as the fulcrum. Consequently, a coaxial high-speed rotation is given to the tip of the wire by the electrical motor. With the conventional weaving method, it is impossible to achieve high-speed weaving at a frequency higher than 10 Hz because of mechanical restrictions. This system easily attains rotating frequencies ranging up to 100 Hz or even higher.

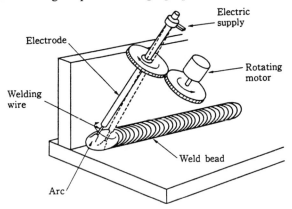

**Fig. 28.1** Princple of the high-speed rotating arc process.

### 28.2.2  Advantage in bead formation

Figure 28.2 shows a comparison of the bead shape with and without rotation. The welding current is 400 A (1.2 mm wire) and the welding speed is 100 cm/min. As is evident from the figure, the conventional process (broken line) produces a considerably convex bead with penetration concentrated at the centre. The rotating process (solid line) produces a flat bead with a smooth surface and less penetration at the centre. This is attributed to the uniform distribution of heat input and pressure of the arc due to the high-speed rotation. This feature is effective for increasing the welding speed.

## 28.3  AUTOMATIC SEAM TRACKING CONTROL BY ARC SENSOR

### 28.3.1  Principle of the arc sensor

The arc sensor is a sensing system designed with an oscillating welding arc in the groove. It can correct the torch position in real time by detecting

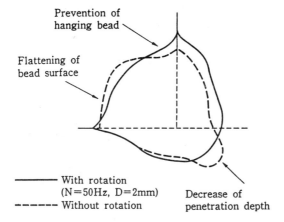

**Fig. 28.2** Improvement of the bead shape using a high-speed rotating arc. 1.2 mm diameter solid wire; $I_a = 400$ A; $v = 100$ cm/min.

deviation between the groove centre and the oscillation centre from the waveform of welding current and/or arc voltage. Because the arc itself serves as a sensor in this system, its features include no need of any special detecting device, simple welding torch assembly and detection of information directly beneath the arc in real time [1].

The automatic seam-tracking performance of the arc sensor is dramatically improved in the high-speed rotating arc process. The principle is shown in Figure 28.3. The arc rotating position (Cf, R, Cr, L) is detected by a

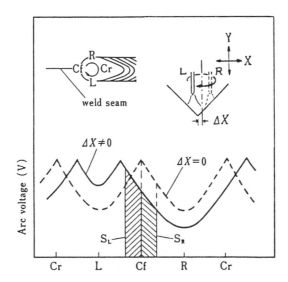

**Fig. 28.3** Principle of the arc sensor.

rotary encoder. The figure shows the basic pattern of the arc voltage waveform correlated to the rotating position of the arc. When the centre of the welding torch is at the centre of the groove ($\Delta X = 0$), as shown by the broken lines, the arc voltage waveform peaks at points Cf, Cr and bottoms out at points R, L and it becomes symmetrical at point Cf in the forward welding direction.

When the torch deviates towards the right (R) side ($\Delta X \neq 0$), the phase of the peak at Cf advances and the waveform becomes asymmetrical with respect to Cf, as shown by the solid lines. Now, by comparing areas (integral values) $S_R$ and $S_L$ of the arc voltage waveforms surrounded by the same phase angle on the right and left sides of point Cf, the deviation of the aiming position of the torch in the x-axis direction is detected. In practical application, since the shape of the molten metal will cause distortion in the waveform at the rear side (Cr) of the rotation, the value of the phase angle is selected to less than 90°. Here, the arc voltage waveform is used for seam-tracking control. However, in principle it is also to use the welding current waveform. As to the torch height direction (y-axis), control is performed at each rotation cycle so that the area of the welding current waveform is constant.

### 28.3.2  Performance of arc sensor with high-speed rotating arc

In order to confirm the detecting ability of the arc sensor in the high-speed rotating arc welding process, the welding current waveforms were observed by changing the arc rotation speed from 1 Hz to 50 Hz in flat fillet welding. The current waveforms thus obtained are shown in Figure 28.4. As is evident

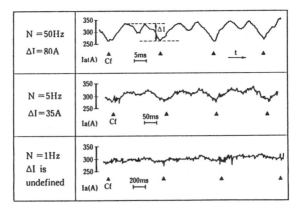

**Fig. 28.4** Influence of rotation speed on welding current waveform. 1.2 mm diameter solid wire; 20% $CO_2$–Ar; electrode wire feed speed, $v_f = 8.92$ m/min; $I_a = 300$ A; arc voltage $E_a = 31$ V; tip–work distance, $E_x = 25$ mm; welding speed, $v = 70$ cm/min; diameter of arc rotation, $D = 3.5$ mm.

from the figure, at one cycle of rotation ($1/N$), each waveform takes the lowest value at point Cf in the forward welding direction, the peaked value at points R and L at both sides, and the mean value at point Cr in the reverse direction. The difference of the value between Cf and Cr can be interpreted as the reflection of the height of the molten pool directly beneath the arc. The amplitude of the waveform or the welding current variable (indicated as $\Delta I$) decreases with the lowering of rotating speed $N$, and at $N = 1$ Hz the waveform shows almost no regularity. Figure 28.5 shows the relationship between the rotating speed and the welding current variable. $\Delta I$ increases with an increase of rotation speed $N$ in the steady state ($N \rightarrow 0$ Hz), $\Delta I = 12$ A, but when $N = 50$ Hz, $\Delta I$ is 80 A, about seven times the value in the steady state.

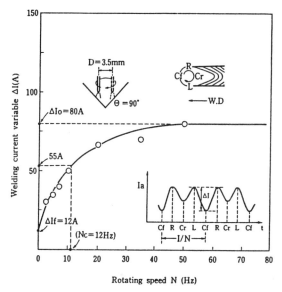

**Fig. 28.5** Relationship between rotation speed and welding current variable. 1.2 mm diameter solid wire; 20% $CO_2$–Ar; $v_f = 8.92$ m/min; $I_a = 300$ A; $E_a = 31$ V; $E_x = 25$ mm; $v = 70$ cm/min; $D = 3.5$ mm.

With the arc sensor of the conventional weaving method, the oscillation frequency is limited to the order of several Hz. However, a frequency of rotation up to 100 Hz is easily obtained with this method. It can be seen that highly precise and responsive automatic seam tracking control is attained with this method thanks to its high detecting ability (welding current variable) as arc sensor and its high sampling speed (rotary weaving frequency).

### 28.3.3 Test results of arc sensor

Figure 28.6 shows an example of test results of automatic seam tracking with a high-speed rotating arc sensor using an articulated arc welding robot.

**Fig. 28.6** Results of automatic seam tracking by arc sensor. 3.2 mm lap joint; $v = 120$ cm/min; $I_a = 300$ A; $N = 50$ Hz; $D = 1.5$ mm.

The specimen used is a lap joint with plate thickness of 3.2 mm and length 500 mm, and its weld seam is curved as shown in the photograph. Although the teaching operation of the robot is performed only at the start and end points of the weld line, it can be seen that automatic seam tracking control has been achieved with a satisfactory bead appearance. It is confirmed that the arc sensor is applicable to a welding speed of max 2.5 m/min (lap joint with plate thickness of 2.3 mm) and a plate thickness of min 1.2 mm lap joints (welding speed 1.5 m/min).

## 28.4   APPLICATION

### 28.4.1   Narrow-gap welding equipment

Figure 28.7 shows the narrow-gap welding equipment in practical application. The work is a large piston cylinder with an outer diameter of 2350 mm and a wall thickness of 275 mm. The welding is carried out continuously

**Fig. 28.7** Narrow-gap welding equipment.

with rotation of the cylinder by a turning roller, so that the arc time ratio is almost 100%.

This process has excellent cost performance. According to actual results, the reduction of the welding cost is about 90% in labour cost and about 70% in welding consumables compared with the conventional method by SAW with V-groove [2].

### 28.4.2    Multi-electrode automatic fillet welding equipment

The multi-electrode automatic fillet welding equipment applied in NKK factories is shown in Figures 28.8 and 28.9. Figure 28.8 shows an eight-

**Fig. 28.8** Eight-electrode line welder for shipbuilding panel line.

**Fig. 28.9** Four-electrode gantry welder for built-up H-beam construction line.

electrode line welder installed in the shipbuilding panel line, and Figure 28.9 shows a four-electrode gantry welder installed in the fabrication line for build-up H beams. Two-electrode tandem fillet welding with high-speed rotating arc process is applied in both installations. Available leg length by one-pass welding is 5–12 mm. Welding conditions at various sizes of leg length are preset in the microcomputer, so those can be automatically selected simply by input of leg length size before welding. High-speed welding of 100 cm/min for a leg length of 7 mm can be achieved in this equipment. All the welding procedure, from initial setting of the welding heads to the work until end removal and return to start position, is carried out fully automatically by a sequence controller, so no manual operation is necessary during welding. This equipment has contributed to the promotion of efficiency in practical use [3], [4].

The high-speed rotating arc process is also applicable to positional welding such as vertical and overhead. Eight-electrode automatic welding equipment for the boiler panel line has been developed and applied as shown in Figure 28.10. This equipment welds tube-fin fillet joints of boiler membrane panels simultaneously from both sides by each of four welding torches in flat and overhead position. Thermal deformation and excess work for turning over the panel are eliminated by this process.

**Fig. 28.10** Eight-electrode welder for boiler panel line.

### 28.4.3    Articulated arc welding robot

Along with extensive applications for arc welding robots, demand for improved functions has been increasing. In particular, great stress has been recently placed on increasing both the high-speed weldability and the sensitivity of arc sensor control in thin plate welding. In response to such demands, a six-axis articulated arc welding robot with a high-speed rotating arc has been developed, as shown in Figure 28.11 [5].

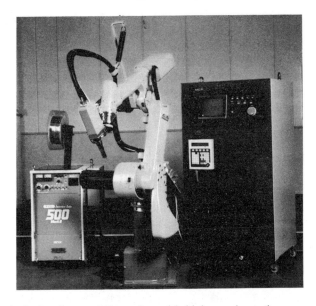

**Fig. 28.11** Articulated arc welding robot with high-speed rotating arc.

**Fig. 28.12** Comparison of bead shape (a) without and (b) with arc rotation. 2.3 mm lap joint; 1.2 mm solid wire; 100% $CO_2$; $v = 250$ cm/min; $N = 50$ Hz; $D = 1.5$ mm.

Figure 28.12 shows a comparison of the cross-sectional bead shape in thin plate welding between this process and the conventional process (without rotation). The welding speed is 250 cm/min, the welding current is 350 A and the plate thickness of the lap joint is 2.3 mm. Because of high current and high-speed welding, the bead shape in the conventional process shows a convex surface and deep penetration to the rear side, and burn-through has occurred. Compared with this, a satisfactory bead shape with a small penetration is obtained in the rotating arc process. This shows that high-speed rotation of the arc is effective for high-speed welding of thin plate. Table 28.1 shows examples of welding conditions in thin plate welding.

**Table 28.1** Examples of welding conditions in thin plate welding

| Plate thickness (mm) | Welding speed (cm/min) | Welding current (A) | Wire diameter (mm) |
|---|---|---|---|
| 3.2 | 220 | 400 | 1.2 |
| 2.3 | 250 | 350 | 1.2 |
| 1.6 | 200 | 250 | 1.2 |
| 1.2 | 150 | 150 | 0.9 |

$CO_2$ 100% shielding gas, $N = 50$Hz, $D = 1.5$ mm

## 28.5  CONCLUSION

1. The precision and responsiveness of the arc sensor has improved dramatically as a result of adopting the high-speed rotating arc welding process. At present, this system is applicable to welding speeds up to 2.5 m/min and plate thicknesses down to 1.2 mm. This is due to the increase of weaving (rotating) speed from the order of several Hz in the conventional process to the order of several tens of Hz.
2. A flat bead with a small penetration depth can be obtained thanks to the high speed rotating of the arc. This characteristic is effective for increase of welding speed and stability of welding quality.
3. This process has been already applied to various practical uses since 1982 with a good performance for welding efficiency and quality.

## REFERENCES

1. Nomura, H. and Sugitani, Y. (1986) Sensing and control of arc welding, in *Advances in Welding Science and Technology, Proc. TWR 86*, pp. 427–34.
2. Nomura, H., Sugitani, Y. and Kobayashi, Y. (1982) Narrow gap MIG welding process with high speed rotating arc. IIW Document IIW-SG212-527-82.
3. Nomura, H., Sugitani, Y., Kobayashi, Y. and Murayama, M. (1986) Development of automatic fillet welding process with high speed rotating arc. IIW Document IIW-SG212-655-86, IIW-X-939-86.
4. Sugitani, Y., Kobayashi, Y. and Murayama, M. (1989) Development of multi-electrode automatic fillet welding equipment with high speed rotating arc, in *Recent Trends in Welding Science and Technology, Proc. TWR 89*, pp. 365–70.
5. Sugitani, Y., Murayama, M. and Yamashita, K. (1990) Development of articulated arc welding robot with high speed rotating arc process, in *Proc. Fifth Int. Symp. of the Japan Welding Society*, pp. 525–30.

# CHAPTER 29

# Groove tracking control by arc-welding current

*Mitsuaki Otoguro*

In V-groove welding in the flat position using gas-shielded arc welding with fine wire by oscillation method, the arc welding current shows a maximum value at both sides of the groove. As the electrode wire approaches the groove wall, the arc welding current gradually increases because the wire extension becomes shorter. As the electrode leaves the groove wall, the arc current decreases rapidly because the arc length is extended. These changes of welding current during welding show the symmetrical butterfly-like arc current-weaving position pattern, and we can detect variations of groove width by detecting the changes of this pattern.

## 29.1 INTRODUCTION

The spread of arc welding robots has entered its second stage under the severe competition between manufacturers for reduction of manufacturing cost and increased sales share. Today's arc welding robots play an important role in industry in cost saving, high productivity and high quality of goods. Arc welding robots have been applied in industrial fields such as automobiles, metals and civil construction which occupy 70% of the demand for the welding robots, where one-pass welding is needed for a thin steel plate.

However, welding robots are now shifting their market to the industries of steel frames, bridges, heavy electric machinery and shipbuilding, where multilayer welding is required for thick plates. In work on thick steel plates, because of the poorer accuracy in processing and attachment of large objects to be welded, welding conditions such as groove line, groove width and wire extension have to be checked and controlled at every pass in order to obtain

*Sensors and Control Systems in Arc Welding.* Edited by Hirokazu Nomura.
Published in 1994 by Chapman & Hall, London. ISBN 0 412 47490 5

sound multilayer welded joints. For this purpose, a preteaching procedure has to be given to an arc welding robot, and this causes inefficiency in the automatic welding operation. Therefore, a non-teaching type of arc welding robot is desired in this field, to simplify the teaching procedure and improve the process accuracy. We have developed a groove-tracking control system applying the arc sensor method in a pure $CO_2$ gas-shielded arc welding process in which continuous multilayer welding in the flat position is operated as a one-pass, one-layer technique.

## 29.2   OUTLINE OF THE SYSTEM

In this system, the $CO_2$ gas-shielded arc welding process is applied to provide suitably flexible operation. The welder consists of three essential components; the $x$ axis for torch weaving and groove line and width tracking control; the $z$ axis for torch movement up and down and wire extension tracking control; and the $y$ axis for torch travel in the direction of the welding line on the guide rail, and bead height tracking control. This basic construction makes the unit small, light and portable.

The controller consists of a 10 Hz low-pass filter to minimize ripple on the welding current signal, an analogue–digital converter, a microprocesser unit and a motor driver for each axis, as shown in Figure 29.1.

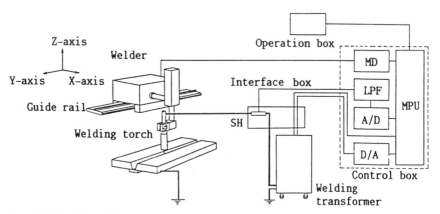

**Fig. 29.1** Outline of the system: MD, motor driver; LPF, low-pass filter; A/D, D/A, converters; MPU, microprocessing unit; SH, shunt.

## 39.3   PRINCIPLE OF THE ARC SENSOR

The concept of the arc sensor in the control system starts with the relationship between welding current value and torch weaving position as shown in Figure 29.2.

Welding current

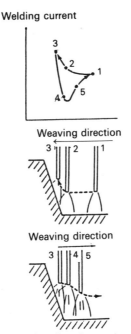

Weaving direction

Weaving direction

**Fig. 29.2** Relationship between weaving direction and welding current at groove wall.

In welding inside a V-groove in the flat position, the welding current value shows a maximum at both ends of the weaving. In this case, the change of arc length is delayed compared with that of the wire extension near the ends. By the time it reaches the ends, the arc length becomes short while the welding current increases.

After reversing the weaving direction, the welding current rapidly decreases according to the growth of the arc length while the current value during forward and reverse weaving changes near both ends. The change of welding current with weaving position makes a symmetrical butterfly-shaped pattern as shown in Figure 29.3. On the basis of the change of this pattern, an aberration of the groove or the welding line can be detected. Figure 29.3 shows that welding is being done in the correct position. $I$ is the average of the welding current.

### 29.3.1 Welding line tracking

When the aberration of a groove occurs to the right or the left, as shown in Figure 29.4, the symmetrical butterfly-shaped pattern becomes distorted, resulting in a difference between areas $A_L$ and $A_R$. The difference $T$ $(t = A_L - A_R)$ is amended in comparison with reference $t$ as follows:

$$\begin{bmatrix} T < -t\text{: Torch shift to left} \\ -t \leqslant T \leqslant t\text{: No change} \\ t < T\text{: Torch shift to right} \end{bmatrix}$$

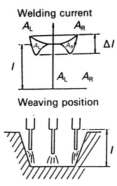

**Fig. 29.3** Relationship between weaving position and welding current with reversal of weaving direction.

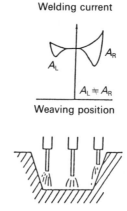

**Fig. 29.4** Welding line tracking.

### 29.3.2   Control of wire extension

When the aberration of a groove occurs up or down, as shown in Figure 29.5, the average current $I$ increases or decreases, resulting in a difference from the reference current $H_{max}$, $H_{min}$. The former is amended in comparison with the latter as follows:

$$\left[\begin{array}{l} I < H_{min}\text{: Torch shift to down} \\ H_{min} \leqslant I \leqslant H_{max}\text{: No change} \\ H_{max} < I\text{: Torch shift to up} \end{array}\right]$$

### 29.3.3   Groove width tracking

When the groove width changes, the difference $\Delta I$ between the current values at the ends of the weaving changes, resulting in a difference between the reference values $W_{max}$ and $W_{min}$ in Figure 29.6. The former is amended in comparison with the latter as shown in Table 29.1,

Welding current

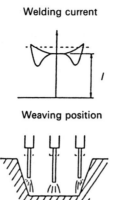

Weaving position

**Fig. 29.5** Control of wire extension.

Welding current

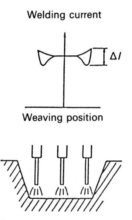

Weaving position

**Fig. 29.6** Groove width tracking.

### 29.3.4 Bead height control

In continuous multilayer welding, the bead has to maintain its height at each pass in order to achieve an equal final pass. According to Figure 29.7, the relationship between bead height and cross-sectional area $M$ in the groove is

$$M = d\left(R + \frac{d \tan \theta}{2}\right) \qquad (29.1)$$

The relation between welding speed $v$ and area $M$ is

$$M = \frac{m}{v} \qquad (29.2)$$

**Table 29.1** Groove width tracking

| $\Delta I < W_{min}$ | $W_{min} \leqslant \Delta I \leqslant W_{max}$ | $W_{max} < \Delta I$ |
|---|---|---|
| Wider weaving | No change | Narrower weaving |
| Later travelling | No change | Earlier travelling |

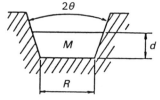

**Fig. 29.7** Groove and cross-sectional area at the bead.

where $m$ is the melting speed of the welding wire per unit time. Furthermore, the relationship between welding speed, beal height $d$ and groove width $R$ is

$$v = \frac{m}{d\left(R + \dfrac{d \tan \theta}{2}\right)} \qquad (29.3)$$

If the groove angle $\theta$ is fixed, the welding speed is obtained from the function of groove width for the maintenance of the present system by

$$v = \frac{a \times cl}{P} \qquad (29.4)$$

where $a$ is the travelling distance per one pulse, $cl$ is the basic pulse frequency, and $P$ is the pulse rate. The welding speed is controlled by the speed control pulse rate $P$. Therefore, according to Equations (29.3) and (29.4), there is a direct proportionality between $P$ and $R$:

$$P = \alpha(R + \beta)$$

where $\alpha = a \times cl \times d/m$, $\beta = d \tan \theta/2$. Accordingly, the pulse rate changes in coupling with the groove width in the groove width tracking control described above while the bead height is fixed by controlling the welding speed as shown in Table 29.1.

### 29.3.5   Method of tracking amendment and memory tracking

The present system permits tracking control in accordance with changes of welding line, groove height and groove width by sensing one signal as three kinds of information. The tracking aberration is amended at every half cycle of weaving from the stored data of one cycle, where amendment is made by a fixed amount at one time. This amendment prevents noises that are not removed through the low-pass filter from causing errors. Furthermore, amendment at every half cycle allows the system to react quickly and accurately. In conventional systems, it has been difficult to detect changes of current value at both ends of weaving because of the shallow groove near the final welding pass. The arc sensor method cannot be applied since a groove is hardly seen, especially at the final pass. In order to resolve these problems, memory tracking is applied to the present system and enables it to track the final phase.

## 29.4  SUMMARY OF AUTOMATIC CONTINUOUS MULTILAYER WELDING

The present system is applied to $CO_2$ gas-shielded multilayer arc welding, under computer control, where continuous multilayer welding is operated with back-and-forth welding for one-pass, one-layer by means of changing the arc targeting point to the change of groove and welding conditions, with an arc-sensing groove-tracking system. As shown in Figure 29.8, multilayer welding is operated by putting crater filler into each end. A perfect arc start has been achieved by applying the process of Figure 29.8 (a)–(e) in order to overcome arc failure resulting from extinguishing the arc at the crater portion.

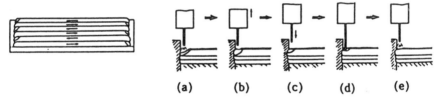

(a)     (b)     (c)     (d)     (e)

**Fig. 29.8** Multilayer method.

## 29.5  RESULTS OF WELDING TEST

### 29.5.1  Welding materials and conditions

The welding materials used were as follows:

Welding wire: Fine wire, Nittetsu YM-26, 1.6 mm diameter
Plates: SM-41B, 32 mm in thickness
Shielding gas: Pure $CO_2$, 25–30 $l$/min
The groove geometry is shown in Figure 29.9, and the welding conditions are listed in Table 29.2.

### 29.5.2  Results of groove tracking

The results are shown in Figure 29.10.

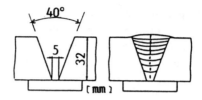

**Fig. 29.9** Groove geometry.

**Table 29.2** Welding conditions

| Pass no. | Welding current (A) | Arc voltage (V) | Welding speed (cm/min) |
|---|---|---|---|
| 1 | 360 | 33 | 35 |
| 2 | 380 | 33 | 30 |
| 3 | 390 | 33 | 25 |
| 4 | 400 | 33 | 25 |
| 5 | 400 | 33 | 25 |
| 6 | 400 | 33 | 25 |
| 7 | 400 | 33 | 25 |
| 8 | 400 | 35 | 20 |
| 9 | 400 | 35 | 20 |

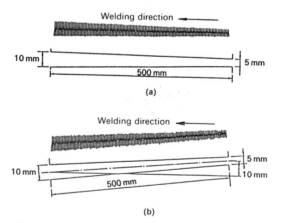

**Fig. 29.10** Results of groove tracking. (a) Groove width tracking: welding length, 500 mm; root gap, start 5 mm, end 10 mm. (b) Groove width and welding line tracking: welding length, 500 mm; root gap, start 5 mm, end 10 mm; difference of line, 10 mm.

## 29.6 CONCLUSION

It has been said that the conventional $CO_2$ gas-shielded arc welding process has difficulty applying the arc sensor method to groove tracking control because of the radical changes of welding current. The present system permits groove tracking control with high accuracy and at high speed by a sampling method under computer control.

Furthermore, the system permits continuous multilayer welding without a complicated teaching procedure as given to conventional robots, so this system shows great promise for providing the sensing technology for intelligent arc-welding robots.

# CHAPTER 30

# Automatic seam tracking and bead height control by arc sensor

*Hirokazu Nomura, Yuji Sugitani*
*and Naohiro Tamaoki*

Automatic seam tracking and welding parameter control technology are necessary for the automation and robotization of arc welding. With regard to welding sensor research, an arc sensor which utilizes the welding arc characteristics is currently being developed. This chapter describes the arc sensor control systems developed by the authors: seam-tracking systems for MAG and TIG weaving welding, and bead height control systems for pulsed MAG and TIG welding by deposition amount control.

## 30.1 INTRODUCTION

Reflecting recent trends, such as the diversification of products and the attempts by new industrial countries to catch up with Japan, the need for automation of welding in heavy industries as well as in shipbuilding has become more pronounced. Such automation can be achieved through various policies: by introducing labour-saving measures; by implementing automated control; by raising efficiency; and by stabilizing weld qualities. To achieve automation of welding, automatic control of seam tracking by the welding torch and of welding parameters (including welding speed, wire feeding rate and welding current) is indispensable. For that purpose, it is necessary to sense the position and form of the groove and the state of the welding zone during welding in real time, rather than relying on the operator's visual judgement.

The authors started developmental studies of arc sensors at an early stage. As a result, an automatic welding system equipped with seam-tracking

*Sensors and Control Systems in Arc Welding.* Edited by Hirokazu Nomura.
Published in 1994 by Chapman & Hall, London. ISBN 0 412 47490 5

sensors have started being deployed in heavy industries. Furthermore, the authors are now developing sensor systems for application to welding parameter control, and these are already in part being put to practical use. This chapter outlines the development of the arc sensor system.

## 30.2  SEAM-TRACKING CONTROL BY ARC SENSOR

The arc sensor is a seam-tracking control system which makes use of the characteristics of the welding arc as a sensor. As shown in Table 30.1, the system has a number of features or practical advantages: it can perform real-time control just under the arc, it does not need any separate sensing device, it is hardly affected by wire bending and arc deviation, and its price is reasonable.

**Table 30.1** Features of the arc sensor

| |
|---|
| Advantages |
|    Real-time control system |
|    No need for sensing device around the torch |
|    Not affected by wire bending or arc deflection |
|    Not affected by misalignment of the carriage or rail |
|    Low cost of the equipment |
| Disadvantages |
|    Need for oscillation of the torch or the arc |
|    Dependent on welding conditions |

### 30.2.1  Control of weaving MAG welding

In consumable-electrode welding, including MAG welding, the load characteristics of the arc and the current–voltage characteristics will change as shown in Figure 30.1, if the distance between the electrode tip and the base metal changes while the wire is being fed at a constant feed rate. The diagram shows the case where a 1.2 mm diameter mild steel wire and a constant voltage power source are used. These characteristics are attributed to the wire melting rate which will change according to the self-control characteristics of arc length that have always been known as the basic characteristics of an arc. Since the constant-voltage power source has a drooping grade of several V per 100 A, with respect to the characteristics shown in the diagram, if, for example, the distance between the tip and the base metal decreases, the welding current increases, and the arc voltage decreases. The current-changing ratio against changes in the distance between tip and base metal (A/mm) increases with an increase in welding current (wire feeding rate) and with a decrease in the wire extension. For example, with reference to the diagram, the current-changing ratio is about 5 A/mm where welding current is 300 A.

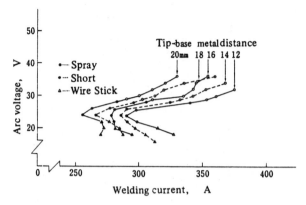

**Fig. 30.1** Arc characteristics in MAG arc welding.

These changes in arc characteristics are used in seam tracking. When the arc is weaved in the groove, the distance between tip and base metal changes in connection with groove angle, which causes changes in the welding current and voltage. Figure 30.2 shows the changes in arc voltage waveform at this time. When the position of the torch weaving centre deviates from the groove centre, the arc voltage waveform becomes asymmetrical with respect to the vertical line. Consequently, when each of the right and left half-cycles of weaving are integrated and areas $S_L$ and $S_R$ are compared, deviation of the torch position can be detected. A block diagram of the seam-tracking control circuit is shown in Figure 30.3.

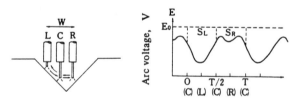

**Fig. 30.2** Principle of arc sensing by integration method.

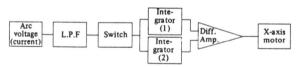

**Fig. 30.3** Block diagram of arc sensor control.

This seam-tracking control has been applied to articulated robots, over-head welding equipment for bridge construction, and so on. Figure 30.4 shows the application of overhead welding equipment for bridge construction.

**Fig. 30.4** View of overhead welding for bridge construction.

### 30.2.2 Control of weaving TIG welding

Since a power source of constant current characteristics is used for the TIG arc, any change in arc length can be detected as a change in arc voltage. In automatic TIG welding, the so-called constant arc length control (AVC) is used. The principle of this control is as follows. The arc length, by virtue of the fact that it is almost proportional to arc voltage, is kept constant by controlling the shifting of the torch in its vertical direction, ensuring that the arc voltage is kept constant. This in turn prevents the electrode from contacting the base metal. The seam-tracking control process is the practical application of this principle.

The principle of the developed control process is shown in Figure 30.5. In the diagram, when constant arc length control is performed while shifting

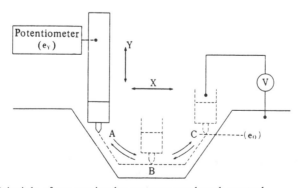

**Fig. 30.5** Principle of arc sensing by constant arc length control.

the electrode in the width direction from point A on the slope in the groove, the tip of the electrode tracks the groove surface while shifting along the path shown by the broken line parallel to the groove surface. The position of the electrode in the direction of torch height at this time is detected using a displacement detector, and when the measured value $(e_Y)$ agrees with the preset reference value $(e_0)$, the shift along the $x$ axis is reversed.

According to this seam-tracking control process, weaving width can be controlled at a correct value at fluctuations in groove width, as well as tracking the groove line. Furthermore, as described later, information on groove width and groove shape can be obtained from the torch weaving displacement. This information can be used for deposition control also. Figure 30.6 shows a block diagram of the control circuit. The upper section is a block diagram of the torch height control circuit and the lower section is that of the $x$ axis control circuit.

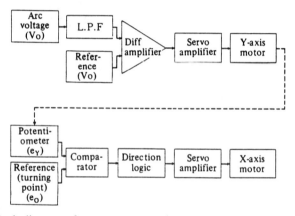

**Fig. 30.6** Block diagram of arc sensor control.

An experimental example of seam-tracking control by this process is shown in Figure 30.7. The experiment was carried out by using a U-shaped groove with a groove face angle of 20° in which the root gap change of about 4 mm between the start point and end point of the groove was given in a taper form and by inclining this groove at 20 mm/300 mm in the travelling direction of the welding carriage. It can be seen from the diagram that tracking control is performed against the deviation of the groove line from the waveform in the $x$ axis and also proper weaving width control is performed against the increase in groove width. Also, it can be seen that at the position with a constant height (point $e_0$), weaving is reversed.

This control process has been applied to automatic orbital TIG welding for a LNG pipeline, and to horizontal TIG welding. Figure 30.8 shows the appearance of the automatic orbital TIG welding.

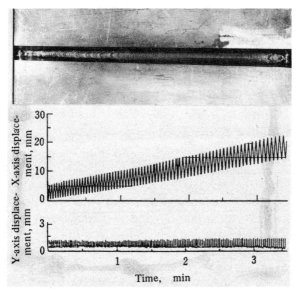

**Fig. 30.7** Results of automatic seam tracking by arc sensor.

**Fig. 30.8** View of automatic TIG welding for LNG pipeline.

## 30.3    BEAD HEIGHT CONTROL BY ARC SENSOR

With the weaving arc sensor using the constant arc length control described in section 30.2.2, it is possible to detect information on groove width and groove sectional area from the weaving displacement waveform of a welding torch, because weaving is reversed at a fixed height based on the base plate level. A deposition amount (bead height) control process is described below. This has been developed using the said function for application to high-current MAG welding using a pulsed current and to all-position TIG welding [2] [3].

### 30.3.1   Control of pulsed MAG welding (control by weaving width)

The seam-tracking control process utilizing the constant arc length control system shown in Figure 30.5 can be applied in the same manner to the MAG welding with a small-diameter consumable electrode. (In this case, the detected wave form is the welding current.) Seam-tracking welding was performed for a variable groove width, butt joint by pulsed MAG welding (wire diameter: 1.2 mm) with an average welding current of 300 A, and the weaving loci obtained are shown classified by groove width in Figure 30.9. Since constant arc length control is performed, the wire extension is kept constant during weaving. Consequently, the weaving locus shown in the diagram may be thought of as the locus of the wire tip, and since the reversing height of weaving, $e_o$, is set at a fixed level from the base metal surface, the weaving width $W$ increases proportionally with an increase of groove width $G$, as shown in the diagram.

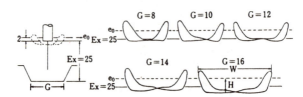

Fig. 30.9 Weaving trace in varied groove width by MAG welding.

Incidentally, the height of the bottom of the weaving locus does not change much despite the fact that the gravity head of the molten metal changes in line with the change in groove width. This phenomenon is characteristic of MAG welding with high current [2]. Figure 30.10 shows

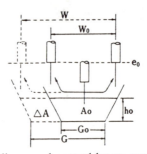

Fig. 30.10 Principle of welding speed control by arc sensor.

the principle of the weld bead height control process in which the welding speed is controlled by obtaining the variables of groove width using only the weaving width as detection parameter based on the said relationship. Let us suppose that a weld bead having bead height $h_o$ and deposited sectional area $A_o$ is formed in a groove having root gap $G_o$, and let the weaving width then be $W_0$. If the groove width has changed to $G$ and the weaving width has become $W$, then the increase in weaving width $(W-W_0)$ is equal to the increase in groove width $(G-G_0)$, under the condition of constant groove angle. To obtain a constant weld bead with a fixed height $h_0$, therefore, deposition has only to be increased by $\Delta A = (W-W_0)h_0$. Consequently, let the initial value of the welding speed be $V_0$ and the wire feed rate be $V_f$; we can obtain the welding speed from the following formula:

$$V = \frac{V_f}{\dfrac{V_f}{V_0} + (W-W_0)h_0}$$

In pulsed MAG welding with a 300 A welding current and a 31 V arc voltage, a root gap $G$ of V-groove joint with groove angle (one side) 25° was made into a taper form having the dimensions 6 mm at the weld starting end and 16 mm at the weld finishing end, and then seam tracking and welding speed control were performed with the result shown in Figure 30.11. The preset inputs were 12.5 m/min for wire feeding rate $V_f$, 5 mm for bead height $h_0$ and 34 cm/min for the initial value of welding speed (i.e. welding speed at the starting end of $G=6$ mm) $V_0$. As can be seen from the diagram, weaving width $W$ increases in proportion to the changes in root gap, and welding speed $V$ is controlled in such a manner that it agrees with the calculated

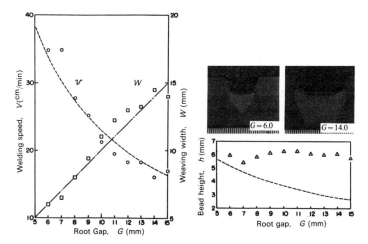

**Fig. 30.11** Result of bead height control by arc sensor in pulsed MAG welding.

value (broken line). Deposition areas were obtained from macroscopic sections at each gap, and bead height was shown as an average height at each section. Although it is about 1 mm higher than the target value 5 mm, it is controlled within the scatter of 0.8 mm. The broken line of bead heights represents the calculated value, based on the assumption that welding speed remained constant at the initial value $V_0$. This control process has been applied to an arc welding robot for building the intersection number of the steel columns of a building and the multipass welding equipment for the diaphragm of a steel column. Figure 30.12 shows the arc welding robot for building core block.

**Fig. 30.12** View of arc welding robot for building core block.

### 30.3.2 Control of TIG welding (control by weaving width and weaving area)

The control process described in section 30.3.1 is based on the premise that there is no change of groove angle or groove bottom height. In high-current pulsed MAG welding, the height of weaving displacement at its bottom part is not correlated with the changes in deposition and groove width due to the strong arc force, so that only the weaving width is used as a detection parameter. In all-position TIG welding with low welding current, the weaving locus of the electrode for performing constant arc length control provides the information on the changes of groove sectional area and deposition sectional area [3].

Figure 30.13 shows the weaving width where the groove width is changed under a constant welding condition. It can be seen from the diagram that the difference between weaving width and groove width is almost constant, regardless of the change of reversing position (point $e_0$) of the weaving. This

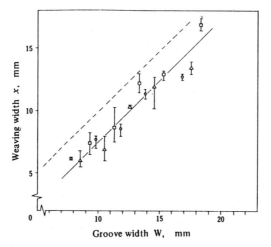

**Fig. 30.13** Relationship between groove width and weaving width. $I = 170$ A; $V_e = 9.0$ V; $\bigcirc$, $H = 2.5$ mm; $\triangle$, $H = 3.5$ mm; $\square$, $H = 4.5$ mm; cap $= 6, 8, 10, 12, 15$ mm; wire diameter $= 1.2$ mm.

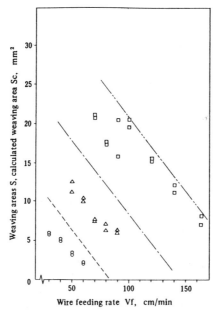

**Fig. 30.14** Relationship between wire feeding rate and weaving area. $I = 170$ A; $V_e = 9.0$ V; $v = 60$ mm/min; $v_x = 500$ mm/min; $\bigcirc$, $---$, $H = 2.5$ min; $\triangle$, $-\cdot-$, $H = 3.5$ mm; $\square$, $-\cdots-$, $H = 4.5$ mm.

value corresponds to the arc length at the weaving end. Figure 30.14 shows the weaving area where the deposition amount changes at three different reversing heights (the height at point $e_0$ measured from the groove root surface). The term 'weaving area' refers to the area enclosed by the weaving locus and the $e_0$ line obtained from each weaving cycle. As is evident from the diagram, the weaving area $S$ decreases linearly with increase of deposition. Furthermore, it is parallel to the calculated value of the weaving area $S_0$, obtained on the assumption that the electrode tip moves with a constant arc length (1 mm) parallel to the bead surface. In other words, the gravity head of the molten metal just under the electrode during weaving is constant regardless of any change in deposition amount. This relationship also holds where the deposition amount is constant and only the groove width is changed.

Using this phenomenon, we have developed a deposition control process in which variations of groove sectional area are detected and the bead height is controlled constantly using the weaving width $W$ and the weaving area $S$. Figure 30.15 shows the principle of this control process. Assume that the initial state of the groove is as in (a). If the groove width changes as in (b), the bead surface also changes as shown in the diagram. As a result, the weaving width and weaving area respectively change from $X_0$ to $X$ and from $S_0$ to $S$.

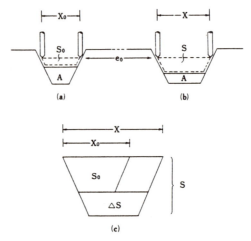

**Fig. 30.15** Principle of deposition amount control.

Consequently, to keep the bead height constant, the deposition sectional area has only to be increased by $\Delta S$, as shown in (c).

Assuming that the height of weaving displacement (reversing height) at the initial state is very small compared with the weaving width, and that deposition control by this process is performed, then the changes in bead height will be also small. Consequently, the value of $\Delta S$ can be approximated from the following equation, which depends only on weaving width and weaving area;

$$S = S - \frac{X}{X_0} \cdot S_0$$

The value of $\Delta S$ is calculated at each weaving cycle, and the wire feeding rate $V_f$ and the welding speed $V$ are controlled by the relation

$$V_f = C(A + \Delta S)V$$

In this equation, $A$ represents the deposition sectional area immediately before the control at each weaving cycle.

Figure 30.16 shows the result of vertical-up TIG welding using this control process. Three-layer welding was performed on a 40° V-grooved

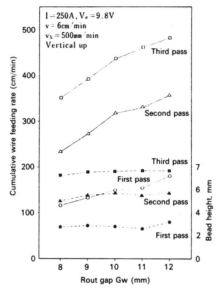

**Fig. 30.16** Result of wire feeding speed control (vertical-up welding).

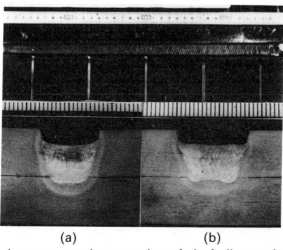

(a)                           (b)

**Fig. 30.17** Bead appearance and macrosections of wire feeding speed control.

joint with varied root gap in taper form in the range 8–12 mm, and the control of wire feeding rate was performed in each layer. As is evident from the diagram, the bead height of each layer is controlled at an almost constant value.

Figure 30.17 shows bead shapes at root gap 8 mm and 12 mm.

Similar results were also obtained for welding speed control at all welding positions including vertical-up and vertical-down positions.

## 30.4   CONCLUSION

Several examples of seam-tracking control and deposition control by the arc sensor control system developed by the authors have been introduced. In the heavy industries and shipbuilding industries, there are strong calls for robotization of welding. To meet this need, it is essential to develop intelligent robots which use sensors effectively. Since the arc sensor does not require any special sensing device, and achieves accurate detection in real time, it may well be a sensor with high applicability to welding robots.

In the future, it will be in particular necessary to develop a welding parameter control system capable of controlling bead shape, deposition amount and depth of penetration by detecting the phenomena of arcs and molten pools.

## REFERENCES

1. Nomura, H., Sugitani, Y., Fujioka, T., Suzuki, Y., Misaura, R. and Arukida, F. (1983) The development of automatic seam tracking with arc sensor. *Nippon Kokan Technical Report (Overseas)*, No. 37.
2. Nomura, H., Sugitani, Y. and Tamaoki, N. Development of multi-pass MAG arc welding system with adaptive control by arc sensor. IIW Document IIW-XII-940-86.
3. Nomura, H., Sugitani, Y, and Suzuki, Y. Automatic real-time bead height control with arc sensor in TIG welding. IIW Document SG212-606-85.

# CHAPTER 31

# Through-the-arc sensing control of welding speed for one-sided welding

*Yasuyoshi Kitazawa*

A development has been made in controlling the soundness of back beads by through-the-arc sensing in GMA one-sided welding with backing support. A correlation exists between the short-circuiting frequency, the distance between the wire and the front edge of the puddle, and the soundness of the back bead. The welding speed is controlled to maintain the short-circuiting frequency at the required level, which corresponds to a distance between the wire and the front edge of the puddle that is optimal for the soundness of the back bead.

## 31.1  INTRODUCTION

One-sided gas-shielded arc welding has been widely used in various fields, including shipbuilding and bridge building. However, in order to ensure high-quality welding the dimensional tolerance for the joint groove (groove angles or root gaps) is narrow, and excellent manufacturing techniques are required to produce uniform and stable back beads (penetration beads). Moreover, in so-called automatic welding, welding operators must monitor the molten pool and the arc state constantly and adjust the welding speed to maintain optimum conditions.

Methods for automatically controlling back beads in one-sided welding with the submerged-arc welding process have been reported [1] [2]. However, these methods require the installation of a sensing device for

*Sensors and Control Systems in Arc Welding*. Edited by Hirokazu Nomura.
Published in 1994 by Chapman & Hall, London. ISBN 0 412 47490 5

detecting the arc light or the flow of plasma at the back of the material to be welded. This makes the operation complex.

To simplify the system, the author and his colleagues have developed a control method to detect the back bead state from the arc's electrical signals, for application in gas-shielded arc one-sided welding with ordinary backing support. This chapter outlines the method.

## 31.2   PRINCIPLE OF THE BACK ARC SENSOR

In a low-current region, which falls within the optimum conditions for one-sided gas-shielded arc welding, the short-circuit transfer mode occurs, in which the arc period and short-circuit period are repeated and droplets are transferred during the short-circuit periods. Figure 31.1 shows the changes in the welding current and voltage in the short-circuit transfer mode; the welding voltage drops to nearly zero during a short-circuit. Although the welding current will continue to increase until there is a short-circuit, it will drop after the short-circuit terminates. Recognizing the occurrence of a short-circuit when the arc voltage $V_s$ falls below the reference voltage $V_{ref}$ will enable the detection of the frequency of short circuits $f$ within a specified time. Figure 31.2 shows the changes in the short-circuit frequency at different welding speeds, and the corresponding back bead appearances. The short-circuit frequency is almost fixed until the welding speed reaches a critical point. Then it is drastically reduced after exceeding that critical point. In addition, the back bead condition improves when the welding

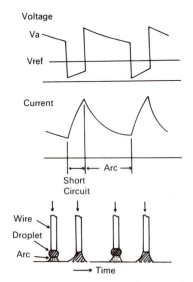

**Fig. 31.1** Schematic illustration of welding voltage and current in short-circuit transfer mode.

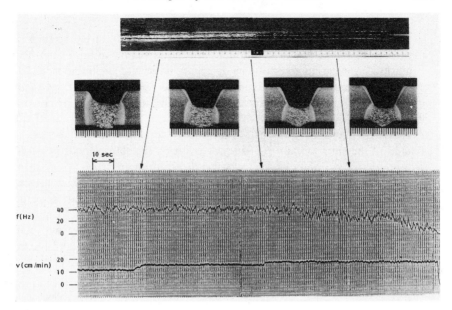

**Fig. 31.2** Change of short-circuit frequency and back bead with welding speed.

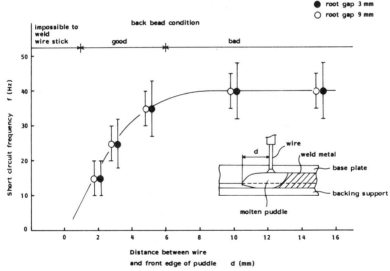

**Fig. 31.3** Correlation between short-circuit frequency and wire position.

speed exceeds its critical point and the short-circuit frequency begins to decrease. If the welding speed increases further, however, the wire will break into the backing support, thus making it impossible for welding to continue.

The correlation between the short circuit frequency $f$ and the distance $d$ between the wire position and the front edge of the molten pool is not related to the root gaps. As the distance between the wire and molten pool front edge decreases, the short-circuit frequency will also decrease, as shown in Figure 31.3. Whereas the distance between the wire and front edge of the molten pool maintains its relation to the back bead condition, the bead condition improves as the distance decreases until wire sticking occurs. The reason for this is that when the wire is placed near the front edge of the molten pool, arc energy can get to the groove back face without being hindered by the molten metal.

Thus whether the back bead condition is good or bad will be closely related to the short-circuit frequency. Measuring the short-circuit frequency by arc voltage will indicate the back bead condition.

Figure 31.4 is a block diagram of the back control method which uses arc sensors. The welding speed $v$ is controlled to keep the short-circuit frequency within the specified $f_{ref}$. If $f > f_{ref}$, $v$ will increase. Since the welding speed increases more than the molten pool moves forward, the wire will come closer to the front edge of the molten pool causing $f$ to decrease and become $f_{ref}$ and finally stop $v$ from increasing. If $f < f_{ref}$, the operation will proceed in the reverse order and $v$ will stop increasing when $f = _{ref}$.

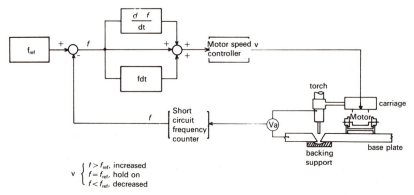

**Fig. 31.4** Block diagram of welding speed control with short-circuit frequency.

## 31.3 APPLICATION RESULTS

Table 31.1 provides an example of the standard welding conditions. In the flat, horizontal and vertical positions these conditions represent practical application.

Figure 31.5 provides an example of the application results, for cases with 3–9 mm tapered root gaps for the $CO_2$ gas-shielded arc welding process in a flat position, 3–7 mm tapered root gaps for the $CO_2$ gas-shielded arc welding process in a horizontal position, and 3–9 mm tapered root gaps for the argon $+ CO_2$ gas-shielded arc welding process in a vertical position. In all cases, the back bead appearance was satisfactory for the different root gaps.

**Table 31.1** Standard welding conditions

| Position | Shield gas | Wire diameter (mm) | Wire extension (mm) | Wire feed rate (m/min) | Welding voltage (V) | Welding current (A) |
|---|---|---|---|---|---|---|
| Flat | $CO_2$ | | | $9 \pm 1$ | $27 \pm 1$ | $220 \pm 20$ |
| | $Ar + CO_2$ | | $25 \pm 5$ | $8 \pm 1$ | $23 \pm 1$ | $220 \pm 20$ |
| Horizontal | $CO_2$ | 1.2 | | $7 \pm 1$ | $25 \pm 1$ | $200 \pm 20$ |
| | $Ar + CO_2$ | | | $7 \pm 1$ | $20 \pm 1$ | $200 \pm 20$ |
| Vertical | $Ar + CO_2$ | | $20 \pm 5$ | $5 \pm 1$ | $19 \sim 20$ | $180 \pm 20$ |

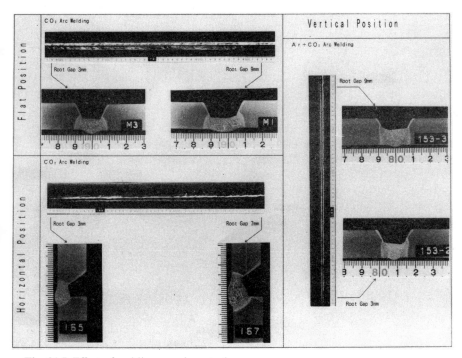

**Fig. 31.5** Effect of welding speed control.

## 31.4 CONCLUSION

In the gas-shielded arc one-side welding process, the short-circuit frequency is related to the distance between the wire position and the front edge of the molten pool. As the distance decreases, the short-circuit frequency decreases.

Controlling the welding speed by using arc sensors will maintain fixed distances between the wire and molten pool front edge even when the root

gaps vary, thereby ensuring that good back beads will always be obtain-
able.

## REFERENCES

1. Nomura H., *et al.* (1982) IIW Document XII-C-032-1982.
2. Nomura H., *et al.* (1985) IIW Document XII-C-107-1985.

# CHAPTER 32

# Some types of wire ground sensors

*Naoki Takeuchi*

The wire ground sensor (wire earth sensor) is the most suitable form of sensor for an arc welding robot because, by sensing the potential difference between the welding wire and the welding workpiece, it is not necessary to mount a separate detector around the welding torch. The welding torch can enter locations on the welding workpiece that are difficult to access. Recently, various versions of this type of sensor have been developed and applied in practice.

## 32.1 INTRODUCTION

If the workpieces to be welded lack fitting accuracy and groove accuracy, a welding robot can sense the weld joint position or gap amount of the workpieces in advance and compute the position or amount accurately to produce good welding results. The wire ground sensor described in this chapter has expanded the application of robot welding for thick-plated construction equipment and steel structures, as well as for large structures. The technological development of wire ground sensors has led to the practical use of various sensing techniques, and robot welding can now be applied to workpieces that were previously considered difficult.

## 32.2 STRUCTURE

Wire ground sensors employ the open-circuit voltage of the welding power source. Groove positions and other positions are sensed with the potential

*Sensors and Control Systems in Arc Welding.* Edited by Hirokazu Nomura.
Published in 1994 by Chapman & Hall, London. ISBN 0 412 47490 5

goes to zero when the wire makes contact with the welding workpiece). Some examples of sensors and their respective sensing methods are described below.

### 32.2.1   Starting-point sensor

This is a sensor to detect the starting point of the welding. If the teaching point (the dotted line in Figure 32.1 showing the fillet welding joint) for which the robot has been programmed differs from the point on the actual workpiece (continuous line), the wire tip in the sensor will sense this and move in the prescribed order by allowing the robot to detect the starting point of the workpiece. The figure provides an example of horizontal fillet welding. The starting-point sensor can also be used for flat fillet welding, single V-grooves, square grooves, single-bevel grooves and lap fillet welding.

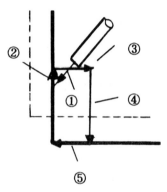

**Fig. 32.1** Starting-point sensor.

### 32.2.2   Three-dimensional sensor

The starting-point sensor is effective for a two-dimensional operation (vertical or horizontal direction orthogonal to the weld line). The three-dimensional sensor can additionally detect differences in the welding directions from the teaching points of the actual workpiece. As illustrated in Figure 32.2, the three-dimensional sensor detects the welding direction in addition to vertical or horizontal.

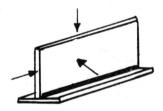

**Fig. 32.2** Three-dimensional sensing.

### 32.2.3 Weld-length sensor

If there is a difference in weld length between the teaching length and the actual workpiece, this weld-length sensor should be used. The sensor detects the starting and ending points of the welding, as shown in Figure 32.3. This method of sensing is very effective for steel structures and other structures with frequently varying weld lengths.

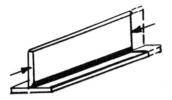

**Fig. 32.3** Weld-length sensing.

### 32.2.4 Multi-point sensor

This is a modified version of the starting-point sensor allowing sensing at any point along the weld line. The multi-point sensor is effective for sensing bent weld lines. Figure 32.4 provides an example. The multi-point sensor exerts its maximum power when used for grooves for which arc sensors are not practical.

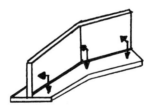

**Fig. 32.4** Multipoint sensing.

### 32.2.5 Intermittent sensor

This is an effective sensing method to weld many short beads on a straight line. The starting-point sensor requires the sensing of each bead; however, with the intermittent sensor, sensing only the first bead is sufficient. This effectively allows for a reduction of search time (Figure 32.5).

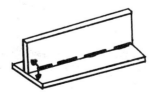

**Fig. 32.5** Intermittent sensing.

### 32.2.6  Stick sensor

The starting-point sensor must enter a groove. Otherwise, the robot ends with an abnormal stop due to an oversensing of the distance. In such a case, if a detectable position which is parallel to the weld line is sensed in advance with a three-dimensional sensor and the distance from the groove position is determined during the teaching time, the groove position can be detected indirectly. However, if there is no such position, detection will be impossible. The stick sensor is a product developed from the image of a human being walking with a stick. With this sensing technique, the robot's torch position moves by a repetition of wire contacting and pulling up from the workpiece. The stick sensor can detect when the robot's torch position is lowered as the wire enters the groove (Figure 32.6).

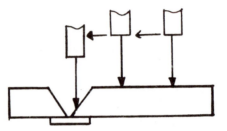

**Fig. 32.6** Stick sensing.

### 32.2.7  Gap sensor

Variations in the groove gap will cause excesses or shortages in the weld metal during normal robot welding. The gap sensor uses the wire to sense the wall sides of a groove in advance and to compare and compute the specified gap size that the robot program should have. It then changes the preset welding speed and weaving width to comply with changes in the groove width. This method of sensing requires a predetermination of the wire size and the groove angle. However, these sizes can be selected and input can be made during the teaching time (Figure 32.7).

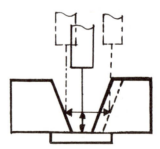

**Fig. 32.7** Gap sensing.

## 32.3 APPLICATION

The following are examples of sensing techniques used individually or in combination.

### 32.3.1 Example 1

This example shows that the welding robot's working ratio and its production efficiency can be improved by combining starting-point sensing and three-dimensional sensing techniques, rather than by having to teach something new for every workpiece, when similarly shaped workpieces are being welded. A and B in Figure 32.8 indicate the teaching workpiece and the actual workpiece respectively.

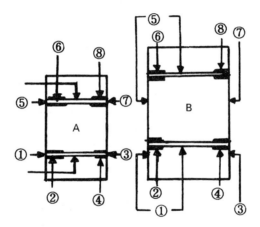

**Fig. 32.8** Example 1.

The figures in the circles under A and B have the same meaning: 1 and 3 are three-dimensional sensing, 2 and 4 are starting-point sensing, while 5 and 7, and 6 and 8 are similarly three-dimensional and starting-point sensing respectively. With these sensing methods, a robot can weld workpiece B by using teaching data A. It is, therefore, not necessary to provide specific teaching for workpiece B. (Only two directions, vertical and horizontal, are used.) This sensing method is effective for welding steel materials and bridge diaphragms.

### 32.3.2 Example 2

For the intermittent beads 1–4 in Figure 32.9, the performance of intermittent sensing from 1 to 2 and from 3 to 4 eliminates the need for starting-point sensing for every bead, thereby enabling large cuts in the search time. This sensing method is effective for welding the loadcarrying platforms of trucks.

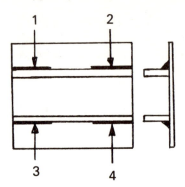

**Fig. 32.9** Example 2.

### 32.3.3   Example 3

Variations in the groove width during butt welding result in excesses or shortages of the weld metal. The thicker the plate thickness is, the greater the excess or lack of weld metal. Figure 32.10 presents a case in which the post connections of a steel box are underlaid by a robot. The gap sensor and starting point sensor detect the groove width and groove centre to perform multilayer welding. These sensors detect the width of the first layer and successive layers, thus forming beads which have the proper width. Figure 32.10, A shows the groove conditions contained by a robot, while B shows the actual groove condition of the workpiece. If the groove width changes from $G=6$ to $G=4$ and to $G=8$, the sectional area will be computed and the variation will be taken into account by increasing both the welding speed and weaving width. The changes in the first layer will be reflected on the second layer and following layers.

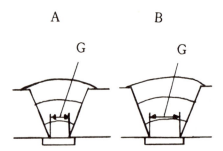

**Fig. 32.10** Example 3.

## 32.4   CONCLUSION

These relatively simple wire ground sensors have enabled robot welding to be applied to workpieces which were considered impractical. However, it should be noted that workpiece and weld line position or gap sensing prior to welding will not take account of movements which occur during welding due to thermal distortion. In part, such movements can be compensated for by interrupting welding and sensing the joint periodically during the robot welding program. This of course may not be possible where long continuous welds are required, and will add further to the operation cycle time already increased by the prewelding search routines. Ideally, then, these sensors are best used in conjunction with a seam-tracking sensor, such as a through-the-arc sensor.

# CHAPTER 33

# Application of a touch sensor to an arc welding robot

*Hisahiro Fukuoka*

Sensing technology is one of the key factors in applying welding robots to large structures. Among the various types of sensor, the touch sensor has the advantage that it does not require any special sensing device around the welding torch because the welding wire is used as a detector. This chapter describes the principle of the touch sensor, and shows several examples of its application. Furthermore, the weaving amplitude control function, which is based on the touch sensor's information, is introduced.

## 33.1  INTRODUCTION

The recent spread of industrial robots has been remarkable against the background of labour shortages attributable to favourable business conditions and workers' shying away from harsh working environments. One of the major technical factors in this spread is the rapid progress in the flexibility of robots achieved by making the most of various sensors, which in turn has made it possible to apply robots in fields where automation used to be impossible.

In arc welding, robotization was first limited to industries where specified machining accuracies of parts were relatively easy to achieve and requirements for welding workmanship itself were not very stringent. In recent years, available sensors have increased, and thus robotization has also been rapidly spreading to industries which require difficult-to-achieve machining accuracies of parts, and high welding workmanship, such as construction machinery.

*Sensors and Control Systems in Arc Welding*. Edited by Hirokazu Nomura. Published in 1994 by Chapman & Hall, London. ISBN 0 412 47490 5

Typical types of sensor for arc welding robots are:

1. arc sensors, which cope with errors in welding parts and thermal distortion by controlling the welding position in real time;
2. touch sensors, which detect contact between the electrode tip and the surface of the welding part by making use of variations in voltage.

Of these two types, the touch sensor system developed by Shin Meiwa is introduced in this chapter. The company succeeded in putting the system into practical use faster than any one else in the industry, and the system, known as the sensorless (SLS) system, has since led the industry.

## 33.2 CONSTRUCTION AND PRINCIPLE

### 33.2.1 Construction of hardware

As shown in Figure 33.1, the hardware comprises:

1. the detector or welding wire itself;
2. a specialized detection power supply, located in the detecting unit, to apply a position-detecting voltage to the wire;
3. a specialized circuit for detecting contact between the base metal and the wire in the form of voltage variation.

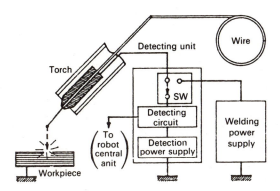

**Fig. 33.1** Hardware of the touch sensor.

Shin Meiwa's touch sensor has two features: the torch is of simple construction because the welding wire itself is used as the detector, and the low-current, high-voltage detecting unit is designed to provide safe detection, and not to be affected by rust and or a thin insulating film on the base-metal surface, associated with objects of welding, for improved detection performance. With these features, the touch sensor is applicable to workpieces of every shape and even those of poor surface condition.

### 33.2.2   Principle and types of correction

The touch sensor is not a tracking sensor, but is of a type that detects errors before welding and accordingly allows modification of the welding program. It comes in two types: the one-point sensor, which detects a single point on the workpiece for error-correction by parallel shifting, and the two-point sensor, which detects two points for correction by rotation.

The principle of error correction is shown in Figure 33.2. Assume that the base metals are located at $S_0$ during teaching. The robot will detect the vertical and the bottom plate, in this order, as shown, and store position $S_0(x_0, z_0)$. If the base metals are displaced to position $S_1$ at playback, the robot will similarly detect position $S_1(x_1, z_1)$. The amount of correction by the sensor is given by

$$E_0(\Delta x, \Delta z) = S_1(x_1, z_1) - S_0(x_0, z_0)$$

and welding point $P_0(x', z')$ will be corrected to $P_1(x, z)$, as given by

$$P_1(x, z) = P_0(x', z') + E_0(\Delta x, \Delta z)$$

Storing this amount of correction, $E_0$, will enable the sensor to correct another teaching point $P_n$ similarly. In other words, the sensor will be able to cope with any workpieces including errors that can be corrected by parallel shifting as shown in Figure 33.2. Although Figure 33.2 shows an error in the $(x, z)$ directions, it is obvious that the touch sensor is applicable to errors in any three-dimensional $(x, y, z)$ directions as well.

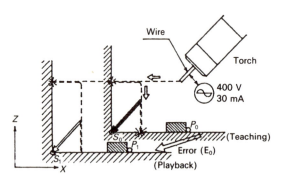

**Fig. 33.2** Principle of error correction.

Touch sensors are divided according to their sensing pattern and sensing direction, as well as according to the correction method.

### (a)   Classification by correction method

1. *One-point sensor*: detects one point and accordingly corrects by parallel shifting.
2. *Two-point sensor*: detects two points and accordingly corrects by rotation.

*(b)   Classification by sensing pattern*

1. *Single-axis sensor* (Figure 33.3a): senses only in a single direction of the vertical or bottom plate.
2. *Fillett sensor* (Figure 33.3b): Senses both of (fillet between) vertical and bottom plates.
3. *Groove sensor* (Figure 33.3c): senses points in the groove, or the intersection between groove walls.

*(c)   Classification by sensing direction*

1. Axial directions of robot: each of $x\pm$, $y\pm$, and $z\pm$ directions.
2. Direction of welding torch (Figure 33.3d): direction of torch projected on

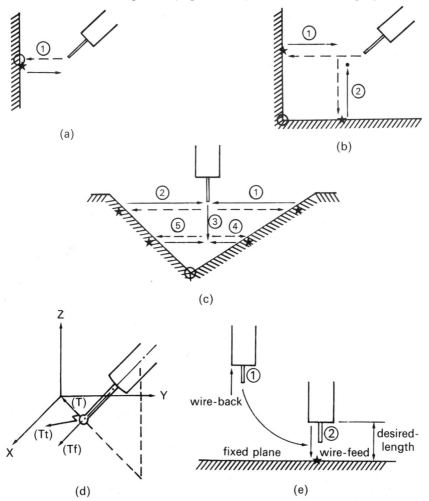

**Fig. 33.3** Typical types of touch sensor: (a) one-point sensor; (b) fillet sensor; (c) groove sensor; (d) sensing direction; (e) extension adjustment.

to xy plane, $T$; direction perpendicular to torch, $T_t$, or direction of torch advance, $T_f$.

For example, the sensor shown in Figure 33.2 is a one-point fillet sensor as it is a fillet sensor to detect the required amount of parallel shift.

Another type of sensor is also available (Figure 33.3e). This automatically adjusts the extension between the torch's contact tip and the base metal in order to eliminate the possibility that variations in the extension may be a factor in sensing error, or have adverse effects on weld beads.

Touch sensors from Shin Meiwa are available in up to 80 types of combinations of (a)–(c) according to the kind of work. Since they are equipped with various devices, such as automatic adjustment of the extension, for high detection accuracy, they will efficiently detect and correct for the amount of error regardless of the kind of work.

## 33.3   EXAMPLES OF APPLICATIONS

As mentioned above, the basic methods of correction by touch sensor are by parallel shifting and by rotation. But combining each amount of correction appropriately will increase the variety of error correction extensively. Some examples of such combinations are described below.

### 33.3.1   Correction by multiple sensing and by multipoint sensing

Correction by multiple sensing refers to the method in which one sensed point is corrected by information obtained by previous sensing. This method will correct for errors that can be removed by parallel shifting plus rotation (three-dimensional errors). In more complex cases, where different kinds of factors are involved, the method will also detect errors in a minimum number of sensings with high precision.

For example, consider a case in which in addition to an error in the installation of the work relative to the fixture, an error in the position of the fixture itself is present, as in Figure 33.4a. In this case, sensing will be carried out in two steps.

1. First, in order to correct for an error in the position of the fixture itself due to, for example, an error in stopping spot on the conveyer, a first sensing will be done on the fixture, or on workpiece 1 directly connected to it.
2. Then, after the amount of correction obtained in the above sensing is reflected in the program, a second sensing will be done on workpiece 2.
3. Finally, by using the amount of correction obtained in step 2 above, the sensor will detect all points on workpiece 2 with high precision.

In correction by multipoint sensing, errors in the work are detected at multiple points beforehand, and each point is corrected by information

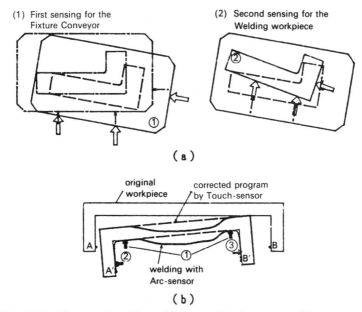

(1) First sensing for the
    Fixture Conveyor

(2) Second sensing for the
    Welding workpiece

( a )

original workpiece

corrected program by Touch-sensor

welding with Arc-sensor

( b )

**Fig. 33.4** (a) Multiple sensing; (b) combining multipoint sensor with arc sensor.

obtained by sensing. This method will allow detection of the positions of the welding start and end points separately, thus leaving no unwelded portion between the start and end points and, further, making it applicable to a series of workpieces that are the same in shape but different in dimensions.

As shown in Figure 33.4b, consider a case in which there is a positional displacement from the original welding line AB to A′B′ plus dimensional differences. In this case, two-point sensor 1 will first detect rotational error. Then, 2 will detect a dimensional error on the A′ side, and 3 will detect a dimensional error on the B′ side. Reflecting in the program the amounts of correction obtained by sensors 1 and 2 for A′, and the amounts of correction obtained by sensors 1 and 3 for B′ will enable continuous welding from A′ to B′. If thermal distortion is also present, the touch sensor may be combined with an arc sensor, not discussed in this paper, to configure an even more powerful sensing system.

### 33.3.2 Groove detection and variation of weaving width

Touch sensors with these functions have been developed for application to workpieces for which robotized or automated welding were impossible because the machining accuracies of their parts or their assembly accuracies were poor and thus positional errors in grooves in them were too large. Examples of such workpieces are the ends of parts forming grooves and bosses in large structures.

The groove detection function is one of detecting a groove accurately and correcting for an error accordingly. As shown in Figure 35.5a, this function is performed in two steps:

1. locating the groove by touch-sensing the surface of the workpiece successively till the specified difference in level is detected;
2. detecting an error in the position of the work in the groove and correcting the welding program accordingly.

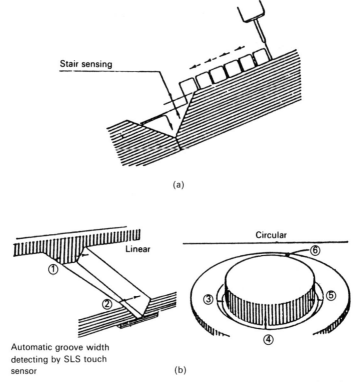

(a)

Automatic groove width
detecting by SLS touch
sensor                (b)

**Fig. 33.5** (a) Groove detection; (b) variable weaving width-changing function.

If the work is inclined, the sensor will provide accurate detection and correction by detecting the inclination of the whole work beforehand and detecting a difference in level in a direction perpendicular to the inclination.

As mentioned above, the sensor is capable of correcting for an error in the position of a groove, but in large structures, groove width in parts to be welded may vary because of assembly variations and these variations in groove width often cannot be ignored. In such a case, uniform weld beads will be obtained by detecting the groove first, and then detecting the groove widths at the welding start and end points, and varying the weaving width and the welding speed according to the detected groove widths (weaving width variation function).

As shown in Figure 33.5b by way of example, proper welding will be achieved by detecting the groove width at two points, 1 and 2, for a linear groove, and at four points 3–6 for a circular groove.

## 33.4  CONCLUSION

This paper introduced the basic principle of the touch sensor system, one of the typical sensors for arc welding robots, and examples of its applications. Today, the touch sensor has found wide use in GTA welding, and flame and plasma cutting robots, etc., as well as in GMA welding robots. Shin Meiwa has taken out many patents, including basic ones, for touch sensor technology, and feels proud that the function of the touch sensor system outdistances that of those from other manufacturers. The company thus feels certain that the touch sensor system will be able to contribute to automation and labour-saving efforts on the part of its users.

The company, building characteristic products such as arc sensors and off-line teaching systems as well as the touch sensor system, will do further research to 'build robots that are easy for anyone to operate and ensure welding of high quality' in order to make contributions to society through technology.

# CHAPTER 34

# Automatic welding for LNG corrugated membranes

*Hirokazu Nomura and Tadashi Fujioka*

Development of an automatic TIG welding system with a contouring sensor has led to improved weld quality and greater efficiency in membrane welding. The contouring sensor is of the contact type, capable of detecting planc inclinations and used for the control of torch movement. In its applications, it was proved that this control system was most suitable for site welding because of its simplicity and stability.

## 34.1  INTRODUCTION

The Technigaz membrane for LNG tanks has excellent structural stability because of its small-pitch orthogonal corrugations. However, these corrugations made it difficult to apply automatic welding to this type of membrane. Since the membrane is thin and is subjected to one-sided single-pass welding, special and delicate skills are required to weld it. Therefore, precise control of the welding torch and reliability for site welding should be considered when an automatic welding system is applied.

NKK developed a compact and lightweight automatic welding machine with a contouring sensor capable of detecting the inclined plane, and applied this automatic welding system on the construction of over ten LNG tanks. This article describes the main features of the control system.

*Sensors and Control Systems in Arc Welding.* Edited by Hirokazu Nomura.
Published in 1994 by Chapman & Hall, London. ISBN 0 412 47490 5

## 34.2 MOVEMENT NECESSARY FOR WELDING TORCH

In order to ensure reliable lap welding of corrugations, the welding torch has to be moved while satisfying the following conditions:

1. The angle of the welding torch to the weld line should be kept constant (position control).
2. The torch tip speed (linear velocity on the weld line) should be kept constant.
3. The distance between torch tip and the base metal (arc length) should be kept constant.
4. To widen the weld bead and stabilize the bead width, regular oscillations should be used, together if possible.

Conditions 1 and 2 are shown schematically in Figure 34.1. Condition 1 indicates that the torch angle should be changed continuously so as to maintain the welding torch substantially perpendicular to the weld line (corrugated surface of membrane). Condition 2 is basic to the welding of thin plates. Condition 3 is of special importance in TIG welding and, as described below, the arc length can be kept constant by proper control of arc voltage. Condition 4 is necessary for ensuring wider tolerance in the exact torch direction and in the throat thickness. To reduce the weight, low-frequency pulsed current was adopted on the developed machine instead of mechanical oscillation.

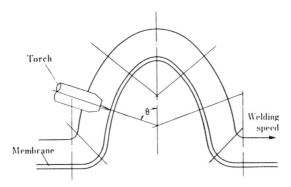

**Fig. 34.1** Torch movement.

## 34.3 CONTROL METHODS ON THE CORRUGATION

### 34.3.1 Contouring sensor

The contouring sensor is of the contact type, having two styluses, and is capable of simultaneously detecting plane inclinations and torch position. Figure 34.2 shows the outlined structure of this sensor. The two styluses are

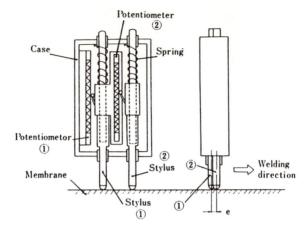

**Fig. 34.2** The contouring sensor.

constantly pushed against the membrane surface and, as shown in the figure, have a slight offset $e$ (1–2 mm in general) in the welding direction. Stylus 1 and potentiometer 1 keep the torch at a constant distance from the membrane. Stylus 2 and potentiometer 2 detect inclination in the welding plane and keep the torch angle to the corrugation surface constant. Figure 34.3 shows the principle of detecting an inclined plane. When the styluses come to an inclined plane, the offset gives rise to a difference in height between them, which is detected by potentiometer 2. By driving the torch rotation motor so that the output voltage of potentiometer 2 becomes zero, the torch can be maintained in a state of zero height difference, i.e. perpendicular to the surface.

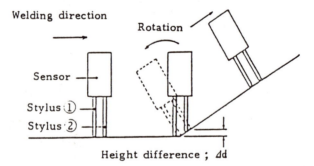

**Fig. 34.3** Principle of detecting an inclined surface.

### 34.3.2   Centre of torch rotation

For a contouring sensor system it is most advantageous for control that the centre of torch rotation is set at the torch tip (arc point). This is because this point is not moved by torch rotation, and arc movement can thus be controlled only by the x and y axes (Figure 34.4). The mechanism of torch rotation is featured by the ring beam or 'bow' as shown in Figure 34.5. While a torch block slides on the bow, the bow itself is rotated by the driving motor. These motions facilitate the full inclination of the torch and prevent the bow from interfering with the corrugation. Moreover, by adopting such a mechanism, one of the motors required for torch rotation could be eliminated, which helped to reduce the total weight of the machine.

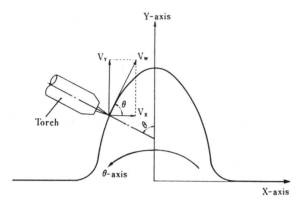

**Fig. 34.4** Speed control around a corrugation.

### 34.3.3 Speed control

To control the torch movement, three basic coordinate axes are defined; the welding direction (x axis); the height direction of the corrugations (y axis); and the direction of torch rotation ($\theta$ axis). As mentioned above, the $\theta$ axis has no relation to welding speed. Only two components, $V_x$ and $V_y$ (see Figure 34.4) can be controlled to govern welding speed. However, it is necessary to keep the torch perpendicular to the corrugated surface. To keep the welding speed $V_w$ at a constant value at all points on the corrugation, $V_x$ and $V_y$ can be determined, so as to satisfy the following equations, as is clear from Figure 34.4:

$$\overrightarrow{V_x} + \overrightarrow{V_y} = \overrightarrow{V_w} \tag{34.1}$$

or

$$V_x = V_w \cos \theta \tag{34.2}$$

$$V_y = V_w \sin \theta \tag{34.3}$$

The value of $V_w$ is set before welding. $\cos \theta$ and $\sin \theta$ can be easily determined by providing a rotation angle detector (for example, a rotary potentiometer, resolver or the like) at the output shaft of the $\theta$ axis motor or the ring gear. By performing the operation of Equations (34.2) and (34.3) using these two values, $V_x$ and $V_y$ can be determined and used for speed

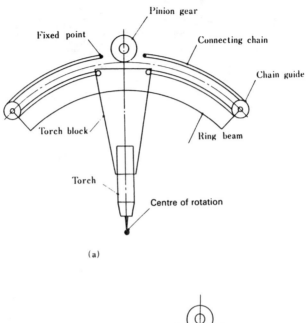

(a)

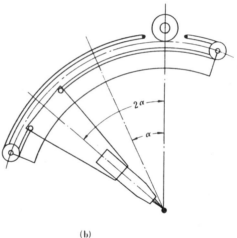

(b)

**Fig. 34.5** Mechanism of torch rotation: (a) before rotation; (b) during rotation.

control. However, in this machine signals for position control are continuously sent to the $y$ axis motor from the sensor. Therefore, addition of the signal corresponding to $V_y$ in Equation (34.1) is likely to render control impossible. When the values of these two signals are equal there is no significant problem but in most cases they are not, and various difficulties result depending upon the characteristics of the control circuit. To avoid this problem, the $y$ axis motor is driven only by a signal from the sensor excluding $V_y$ in Equation (34.3), thus controlling in the simplest way. Even using this method, if the profiling sensitivity of the sensor is high enough, $V_y$ is substantially equal to the value in Equation (34.3), and the welding speed

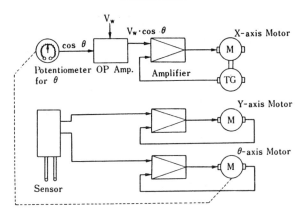

**Fig. 34.6** System for speed control.

can be kept nearly at a constant value. The speed control method is illustrated in Figure 34.6 in block diagram format.

## 34.4  APPLICATION TO SITE WELDING

The fluctuation of the welding speed is caused by various errors in the control system. The first application indicated that the fluctuation ratio on the corrugation was within $\pm 5\%$, and it became larger especially at the foot and top portion of the corrugation. However, it was not found that welding defects corresponded to the fluctuation ratio. Rather, it was because the poor fitting of the membrane and rapid change of the welding position had a great influence on the welding results. As a result of this analysis, to reduce welding defects, it was concluded that smooth torch movement is more important and essential if the speed variation is below allowable range (10%).

Welding parameters for site welding were selected as given in Table 34.1. To achieve high welding quality and efficiency, the welding speed on a flat section is set at about 40% higher value than on a corrugation. Welding parameters are automatically and alternately changed by detecting corrugations. Figure 34.7 shows a macro section of the weld obtained by automatic welding.

## 34.5  CONCLUSION

To improve the reliability and efficiency of the welding of Technigaz-type membranes, a light and compact automatic welding machine was developed and applied to site welding. The machines developed have already been applied to the membrane welding of more than 10 000 m of weld length. These results showed that the torch control system on the corrugation by

**Table 34.1** Welding parameters

| Item | Flat section | Corrugation |
|---|---|---|
| Kind of joint | 1.4 mm'/1.4 mm 'lap joint | |
| Welding process | Low-frequency pulsed TIG | |
| Direction of torch | | |
| Electrode | 2% thoriated tungsten (1.6 mm diameter) | |
| Arc length | About 0.5 mm | |
| Argon gas flow rate | 8 litre/min | |
| Welding current | | |
| Pulse frequency | 3.5–4.0 Hz | 2.7–3.0 Hz |
| Welding speed | 160–180 mm/min | 120–130 mm/min |

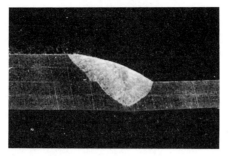

**Fig. 34.7** Macrosection of automatic lap weld.

the contouring sensor had a high reliability and stability. Also, successful applications of automatic welding contributed greatly to the safety of the LNG membrane system.

## REFERENCES

1.  Nomura, H. *et al.* (1979) Development of automatic welding equipment for the corrugated membrane of the LNG tank. *Nippon Kokan Technical Report* (*Overseas*) No. 27.
2.  Nomura, H., Fujioka, T., Wakamatsu, M. and Saito, K. (1982) Automatic welding of the corrugated membrane of an LNG tank. *Metal Construction*, July, 391–5.

# CHAPTER 35

# Welding process monitoring by arc sound

*Masami Futamata*

The sound emitted during the arc welding process is closely related to welding phenomena and provides useful information for monitoring and controlling the welding process. This chapter introduces two examples and examines the effectiveness of monitoring arc sound: one in which holes and steps on base metal were detected, and another in which humping beads and burn-through were detected.

## 35.1 INTRODUCTION

Although obtaining information concerning welding phenomena from the welding process itself is significant to engineering, few effective and practical methods have been developed to obtain such information. One reason is the technological difficulties involved. The heat, light, spattering, fume, general noise, and the electric and magnetic noises emitted by the arc make it difficult to apply conventional measurement technologies to gather information around the area where the arc is generated [1].

The sound generated during arc welding is one of the disturbances to the detection of information during welding. However, this sound is closely related to welding phenomena and can be a source of information for understanding the welding process [2]–[6].

This chapter introduces some practical examples in which arc sound was use effectively to detect phenomena that might cause welding defects, and also describes the effectiveness of monitoring the welding process by arc sound.

*Sensors and Control Systems in Arc Welding.* Edited by Hirokazu Nomura.
Published in 1994 by Chapman & Hall, London. ISBN 0 412 47490 5

## 35.2  EXPERIMENT PRINCIPLES

Generation of arc sound means that the welding phenomena have changed pressure in the arc area where atmospheric pressure is at the equilibrium point. Therefore, arc sound must contain all information regarding the phenomena.

A possible problem when using a sound signal as an information source is the S/N ratio quality. While the steady-state noise level (background noise) is 85 dB(A) at its maximum, the sound pressure level of arc sound is approximately 90 dB in $CO_2$ arc welding (300 A), and 97 dB in MIG arc welding (300 A) [2]. Arc sound is thus higher than the background noise level, and in general the S/N ratio is of good quality. (If the sound subjected to measurement is greater than the background noise by more than 10 dB, the influence of background noise may be ignored.) If the S/N ratio is not of good enough quality, it may be improved by such means as shutting out the background noise, using a unidirectional microphone or a signal-processing algorithm. Being a non-contact and lightweight sensor, an appropriate microphone with any desired sensitivity and frequency range can be easily selected.

The characteristics of arc sound are:

1. waveform;
2. power (all bands, optional bands, frequency load, etc.);
3. frequency spectrum (all bands, optional bands, etc.);
4. autocorrelation function and cross-correlation function;
5. periodicity [2] [3].

In this paper, sound pressure levels of all bands are used.

Sound pressure level (SPL) is given by the equation

$$SPL = 20 \log_{10} \frac{P_e}{P_{e0}} \tag{35.1}$$

$P_e$ represents the effective sound pressure value of the subject being measured, and $P_{e0}$ represents the reference effective sound pressure value, $2 \times 10^{-5}$ Pa. It is one of the most basic physical quantities that expresses the power of the sound, and its relation with power level (PWL) is represented by the equation

$$SPL = PWL - 20 \log_{10} r - 11 (dB) \tag{35.2}$$

where $r$ represents the distance between the sound source and the measurement point.

Figure 35.1 shows a block diagram of the detection system. A microphone was set 400 mm (120 mm for TIG arc welding) from the arc generation site, at a holding angle of 45°. The levels of significant deviation of signals from steady-state arc sound level were set at greater than 3 dB in $CO_2$ and MIG arc welding, and at greater than 1 dB in TIG arc welding.

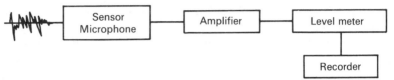

Fig. 35.1 Block diagram of arc sound detection system.


## 35.3   RESULTS

Figure 35.2 shows an example of $CO_2$ arc welding with drilled holes in the base metal. Correspondence is indicated between the change in sound pressure level and the change in bead form over time at the area of the drilled holes.

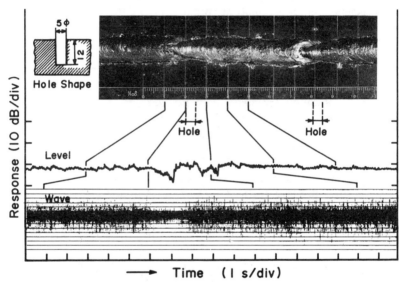

**Fig. 35.2** Example of detection of a hole in $CO_2$ arc welding.

For a hole 5 mm in diameter and 12 mm in depth, the sound pressure level change can be recognized from slightly in front of the hole to approximately 10 mm behind the hole, and the fluctuation range reaches a maximum of 8 dB. This value is significantly greater than the fluctuation when a normal bead is formed, 1 or 2 dB. For TIG arc welding with drilled holes 3.5 mm in diameter, 5 mm in diameter, and 9 mm in diameter (all 12 mm in depth), the observed changes were 2.5 dB, 3.5 dB and 5 dB respectively. Since the level hardly fluctuates in TIG arc welding unless the electrode shape or arc length changes, these values represent good detection signals. The minimum detectable hole was approximately 2.5 mm in diameter for $CO_2$ arc welding, and approximately 1 mm in TIG arc welding.

Figures 35.3 and 35.4 show examples of steps on the base metal. The step height is 20 mm in Figure 35.3, and 8 mm in Figure 35.4. There are two welding directions: (a) from high to low (step down) and (b) from low to high (step up).

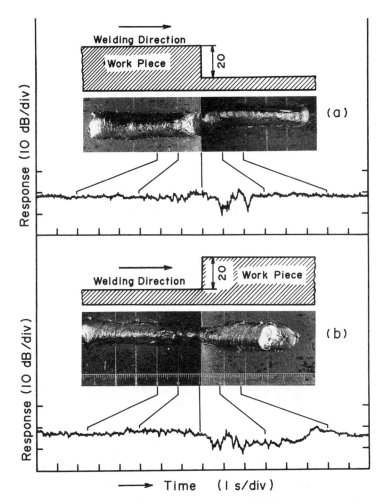

**Fig. 35.3** Examples of steps detected in $CO_2$ arc welding: (a) step down; (b) step up. Welding current, 200 A; arc voltage, 20 V; welding speed, 20 cm/min.

In $CO_2$ arc welding example, the level fluctuation in the range including the step was a maximum of 12 dB at a step up, and a maximum of 13 dB at a step down. The range in which level fluctuation can be easily recognized as being influenced by step is slightly greater in step-up cases.

In pulsed TIG arc welding, however, the level fluctuation is approximately 7 dB in both step-up and step-down cases. The change over time is approximately the same, except that the rising and falling slopes are

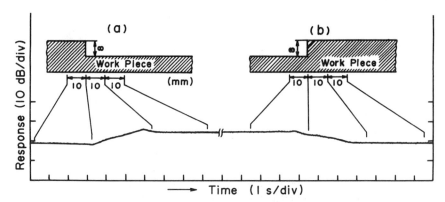

**Fig. 35.4** Examples of steps detected in TIG arc welding: (a) step down; (b) step up. Current waveform, rectangular; $I_p = 150$ A; $I_b = 50$ A; $f_p = 50$ Hz; welding speed, 20 cm/min.

reversed. Although the level fluctuation is caused by a change in arc length, the change does not step up or down, because part of a leg of the arc gradually stretches out parallel to the vertical surface as the torch is moved. These examples resemble the case for bevelling, where the arc moves from inside to outside the bevel because of an error in aim. Arc sound may be used routinely in bevelling.

Figure 35.5 shows an example of humping bead formation detected by arc sound. At each hump, the sound pressure level fluctuates by approximately 1.5 dB. Under the conditions of fixed electric current, voltage and electrode-tip angle, humping beads are generated by increasing welding speed greatly,

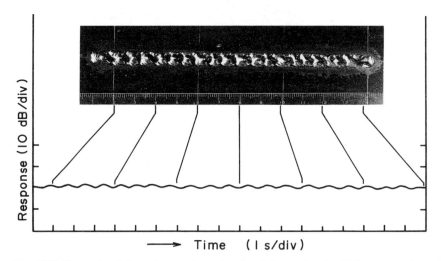

**Fig. 35.5** Example of detection of humping bead formation in TIG arc welding. Current waveform, sine; peak current, $I_p = 200$ A; base current, $I_b = 150$ A; pulse frequency, $f_p = 200$ Hz; welding speed, 40 cm/min.

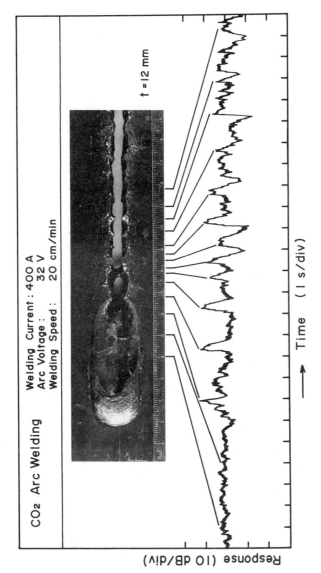

**Fig. 35.6** Example of burn-through detected in $CO_2$ arc welding.

and its mechanism is understood as follows [7]. In fast welding, metal in the molten pool, originally small in quantity, is blown by a plasma current and is absorbed into the bead already formed behind the arc by cohesion force. As a result, the solid metal surface is exposed directly under the arc and the legs of the arc move irregularly, causing the molten pool to move back and forth to form jagged beads. The phenomena of molten metal in the area directly under the arc moving periodically through the side of the arc to the back, and that of a part of a leg of the arc expanding backwards as molten metal moves, have been observed. Therefore, the level fluctuation associated with humping generation is caused mainly by a change in arc length as it expands. Also, the level fluctuation associated with humping generation does not depend on pulse waveform, and it remains approximately in the 1–1.5 dB range for any pulse waveform.

Figure 35.6 shows an example of burn-through detected by arc sound. The level fluctuation associated with burn-through is extremely high, reaching a maximum of 23 dB. Therefore, it may be easily detected by arc sound. The change in levels over time and change in bead shape were observed comparatively, revealing that level peaks are generated periodically and correspond to notch-shaped through-holes. Generated burn-through is generally observed periodically in low-speed welding. In this case, bead shape is a chain of notch-shaped through-holes with the appearance of a cut and containing almost no weld metal in most cases. Notching is a phenomenon that can be generated both in oxygen cutting and in arc cutting, and its detection is an important consideration for quality control of the cutting surface. Arc sound may be a significant information source in these cases, too.

In addition to the phenomena discussed in the above examples, changes in bevel shape and arc length, the existence of a shielded gas supply, and change in transferring form of droplets can also be detected by arc sound [3] [6].

## 35.4   CONCLUSION

Whether all phenomena in the welding process are subject to detection by arc sound and whether differentiation among them is possible are questions whose resolutions require further examination. However, as the above examples demonstrate, arc sound can be a significant and effective monitoring method. Monitoring is possible with a simple system as shown in Figure 35.1, but it is recommended that the system be set in a single processor or be programmed in a minicomputer when introduced into the production line.

## REFERENCES

1.  Arata, Y. and Inoue, K. (1974) Basic problems on automatic control of arc welding. *Journal of Ultra-High Temperature Study*, **11**(3), pp. 1–129.

2. Arata, Y., Maruo, H., Inoue, K., Futamata, M. and Toh, T. (1978) Effects of welding method and welding condition on welding arc sound. *Journal of the Japan Welding Society*, **47**(7), pp. 36–44.

3. Arata, Y., Maruo, H., Inoue, K., Futamata, M. and Toh, T. (1979). Evaluation by hearing acuity and some characteristics of arc sound: Investigation on welding are sound (report 2). *Journal of the Japan Welding Society*, **48**(5), pp. 50–55.

4. Futamata, M., Arata, Y., Maruo, H., Inoue, K. and Toh, T. (1980) Effects of current waveforms on TIG welding arc sound: Investigation on welding arc sound (report 3). *Journal of the Japan Welding Society*, **49**(1), pp. 39–45.

5. Futamata, M., Toh, T., Inoue, K., Maruo, H. and Arata, Y. (1983) Permissible time of arc sound exposure from standpoint of conservation of hearing acuity: Investigation on welding arc sound (report 6). *Quarterly Journal of the Japan Welding Society.* **1**(2), pp. 193–8.

6. Futamata, M., Toh, T., Inoue, K., Maruo, H. and Arata, Y. (1983). Application of arc sound for detection of welding process: Investigation on welding arc sound (report 7). *Quarterly Journal of the Japan Welding Society*, **1**(3), pp. 3–8.

7. Ando, K. and Hasegawa, M. (1967) *Arc Welding Phenomenon*, Sanpo Publishing, p. 304.

# CHAPTER 36

# A vibrating reed sensor for tracing the centre of narrow grooves

*Katsuyoshi Hori, Yoshihide Kondo,*
*Kazuki Kusano and Hiroyasu Enomoto*

This chapter describes a sensor that utilizes the resonance phenomena of a vibrating spring reed. When this reed is inserted in a narrow groove, the tip of the reed taps on the side walls. The phase of vibration and the tapping sounds produced depend on the position of the reed in relation to the groove walls. It is possible to detect derivations from the centre line of the groove with a dead band of $\pm 0.3$ mm. The vibrating reed does not disturb observation of the arc. The sensor has been applied to torch position control for narrow-gap welding in which the groove measured from 6 to 30 mm in width and from 5 to 180 mm in depth, and can operate at temperatures up to 300°C.

## 36.1  INTRODUCTION

Groove-tracing sensors for narrow-gap welding of thick plates have to trace the bottom of weld grooves that are narrow, deep and hot. For this purpose, the authors have developed a new sensor based on a resonance phenomenon of a cantilever flat spring.

The sensor has been applied to the narrow gap GMA welding [1] of components for thermal power plants, nuclear power plants and chemical plants, in which the groove widths measure from 6 to 30 mm and the groove depths measure from 5 to 180 mm.

*Sensors and Control Systems in Arc Welding.* Edited by Hirokazu Nomura.
Published in 1994 by Chapman & Hall, London. ISBN 0 412 47490 5

## 36.2 PRINCIPLE AND STRUCTURE

### 36.2.1 Principle

Figure 36.1 shows the principle of the vibrating reed sensor. The spring reed in resonance vibrates symmetrically as shown in (a). When the amplitude of the vibrations of the reed tip are constricted on one side through contact with an object (b), the amplitude of the vibrating reed on the other side decreases to the same amplitude by virtue of the reed's symmetrical characteristics in resonance.

When the vibrating reed is in the weld groove, the tip of the reed taps both sides of the groove when the centre of vibration is in the centre of the groove (d), and taps only the nearer side when the centre of vibration deviates from the groove centre as in (c) or (e).

By using a piezoelectric (PZT) transducer to detect the tapping sounds, five vibration states can be distinguished: right side tapping, tapping on both sides, left side tapping, vibration without contact, and continuous contact.

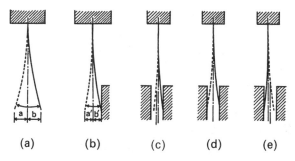

(a)  (b)  (c)  (d)  (e)

**Fig. 36.1** Principle of vibrating reed sensor: (a) $a = b$; (b) $a' = b'$; (c) right; (d) both; (e) left.

### 36.2.2 Structure of the sensor

Figure 36.2 shows the structure of the sensor. The vibrating reed is made of a heat-resistant and ferromagnetic spring material (PSW-2) and its typical dimensions are $0.7 \times 10 \times 160$ mm. As the sensor is cooled by water, it can operate with plates at preheating temperatures up to 300°C.

One end of the sliding base of the sensor is mounted on the support of the welding torch. The tip of the reed is positioned about 30 mm from the tip of the torch, as shown in Figure 36.3. The sensor is set by adjusting the sliding base so that the reed taps both sides of the groove when the welding torch is central.

Figure 36.4 shows the block diagram of the sensing system circuitry. During welding, the sensor position is controlled by feedback through the torch position control unit to maintain this double-sided tapping.

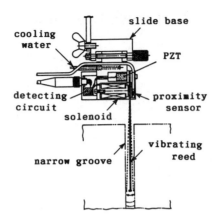

**Fig. 36.2** Structure of sensor.

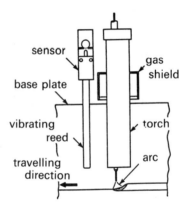

**Fig. 36.3** Setting of sensor and torch.

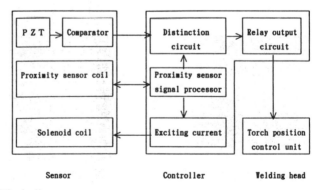

**Fig. 36.4** Block diagram.

### 36.2.3 Excitation mechanism

The reed is attracted periodically to the solenoid in the sensor body, which is automatically tuned to resonate. The resonance frequency is approximately 30 Hz.

When the reed begins to vibrate, it is forced to vibrate by a current which has a frequency that is preset slightly lower than the resonant frequency. After moving slightly, the reed position is separated from the vibrating centre by the proximity sensor. The solenoid is energized to attract the reed while it is on the far side (to the solenoid) half of the vibrating cycle, and not attract the reed while it is in the near half. By adopting this self-induced vibration circuit, the changes of resonance frequency induced by the attachment of spatters to the reed, etc., do not become serious.

The proximity sensor can detect the groove widths by detecting the amplitudes of the vibration during the double-sided tapping.

### 36.2.4 Detection of the tapping sounds

The tapping sounds are detected by the PZT element set near the anchored side of the vibrating reed. The resonant frequency of the PZT element used is 100 kHz. As the main signal components of the tapping sounds detected by the PZT are in the range 20–30 kHz, the mechanical noises from its surroundings are rejected by a 5 kHz high-pass filter.

Figure 36.5 shows the double-sided tapping sound signals after they have passed through the high-pass filter, and the threshold levels of the comparator and the outputs from the sensor. The side walls tested were mild steel surfaces or vinyl taped on to the mild steel.

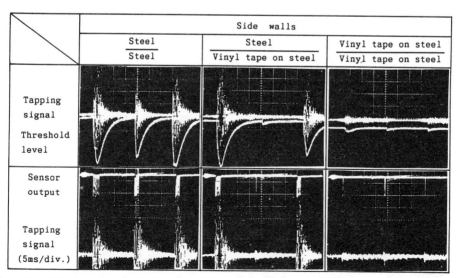

| | Side walls | | |
|---|---|---|---|
| | Steel / Steel | Steel / Vinyl tape on steel | Vinyl tape on steel / Vinyl tape on steel |
| Tapping signal<br>Threshold level | | | |
| Sensor output<br><br>Tapping signal<br>(5ms/div.) | | | |

**Fig. 36.5** Tapping sound signals and output signals from sensor.

By changing the threshold levels, which appear to be enveloped wave forms of the tapping sound signals, automatic gain control is performed. The threshold levels are made by filtering the output from the comparator itself. In spite of the amplitude of tapping sound signals, which were radically different, the output signals from the sensor correspond to each tapping sound as shown in Figure 36.5.

### 36.2.5 Distinction of tapping conditions

The output signals, which are similar to 'chatter' signal, are converted into a one-pulse signal that corresponds to the tapping sounds.

Figure 36.6 shows signals in the distinction circuit. From the relationship between the output signals of the proximity sensor (upper waveforms) and the pulses of the tapping signals (lower waveforms), the sounds are distinguished so that it can be recognized whether each side of the wall is tapped or not.

The signals are output as relay signals, by which the sliding base for the welding torch is driven to position the torch in the groove centre.

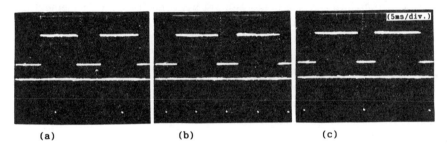

(a)                    (b)                    (c)

**Fig. 36.6** Distinction of tapping sides: (a) right side; (b) both sides; (c) left side.

### 36.3   APPLICATION RESULTS

The sensor and the controller are shown in Figure 36.7. Insertion of the vibrating reed into the groove does not disturb observation of the arc because it vibrates at approximately 30 Hz.

Figure 36.8 shows the relationship between the groove width and deviations from the groove centre in which double-sided tapping is maintained. The accuracy of detecting deviations from the groove centre depends on various factors, such as the size of the reed, the groove width, additional energy during vibration and energy absorbed by the tapped surfaces.

Appropriate dead bands are necessary to prevent the torch from 'hunting' in the groove.

Figure 36.9 shows the practical application of the sensor to circumferential seam welding of a nuclear reactor pressure vessel (6.5 m in diameter and

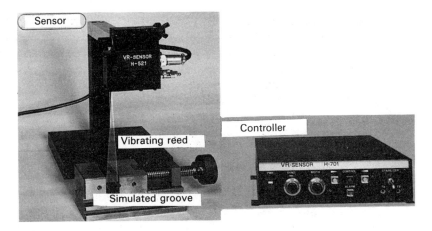

**Fig. 36.7** Sensor and controller.

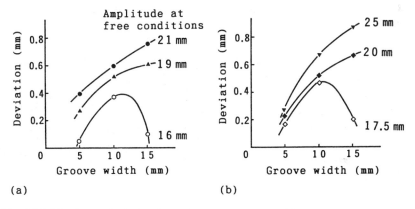

**Fig. 36.8** Maximum deviation from the groove centre while maintaining a double-sided tapping state: (a) reed A, $0.7 \times 10 \times 160$ mm; (b) reed B, $0.3 \times 10 \times 100$ mm.

165 mm in thickness). In this case, the groove widths were between 9 mm and 12 mm, and the vessel was preheated to between 200°C and 300°C during welding. The dead band of the sensor used was $\pm 0.3$ mm. Spatter and fumes from the GMA were not serious over long durations of usage.

## 36.4 CONCLUSION

A sensor, based on a completely new principle, has been developed for tracing the centre of narrow grooves. It applied to narrow-gap welding with groove widths from 5 to 30 mm and depths are from 5 to 180 mm. The sensor can be employed in environments preheated up to 300°C.

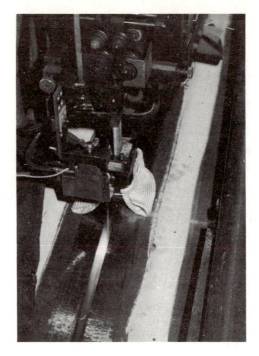

**Fig. 36.9** Application of the sensor to narrow-gap welding of a nuclear reactor pressure vessel.

The sensor has the ability to detect groove widths, but this has not yet been sufficiently developed for practical use.

## REFERENCES

.1  Kawahara, M. and Asano, I. (1986) BHK type narrow gap GMA welding process, in *Narrow Gap Welding – The State of the Art in Japan*, Technical Commission on Welding Process, Japan Welding Society, pp. 39–45.

# CHAPTER 37

# Development of an ultra-heat-resistant electromagnetic sensing system for automatic tracking of welding joints

*Kazuhiko Wakamatsu and Yasuo Kondo*

An ultra-heat-resistant electromagnetic sensing system has been developed and successfully applied to an arc welding robot and an automatic welding machine by continuously providing sensors in the vicinity of the welding arc. This sensing system, based principally on the non-contact, eddy current, gap-detecting process, is designed to be applicable to sensing the location of the welding joint in real time by development of the following functions: (a) Heat-resistant and anti-environmental miniature sensor tip; (b) effective compensation technique for temperature drift of detecting signal; (c) wide range linearization technique of detecting signal; (d) typical specification of the sensing system, sensor tip: 14 mm diameter by 12 mm, fine ceramic covered, coreless, single solenoid type, temperature compensation range $-30-+500°C$, $\pm 2\%$ real scale, linearization range 0–9 mm, $\pm 0.1$ mm.

## 37.1 INTRODUCTION

A welding system has to have the sensor tip placed as close to the welding arc as possible to ensure accurate tracking of the welding joint. This means

*Sensors and Control Systems in Arc Welding.* Edited by Hirokazu Nomura.
Published in 1994 by Chapman & Hall, London. ISBN 0 412 47490 5

that the tip of the sensor is affected by high temperature gas, fumes, spatter, strong magnetic and electric fields, sparks and noise. As well as hindering the performance of the sensor, these effects will also shorten its service life.

In our research, we have successfully developed a system capable of responding to the above demands. It comprises an ultra-heat-resistant, environment-proof and miniature electromagnetic sensor tip coupled with a highly accurate temperature-compensation circuit and a linearization circuit for continuous application to actual machines, i.e, automatic arc welding systems and/or arc welding robots. The test results are summarized below.

## 37.2   CONFIGURATION OF THE SYSTEM

The sensing system is based on a non-contact, high-frequency electromagnetic eddy current displacement meter. In developing the system, the principal aims were as follows. The principal points for this development are as follows.

1. The sensor tip must have uniform detecting sensitivity over a wide range of temperature. The sensor must be as small as possible, and must maintain accurate detection performance in adverse environments.
2. The system must be insensitive to small fluctuations such as spatter deposited on the surface of the material to be welded.
3. The output of the sensor must be linear, with adequate compensation for output drift covered by temperature fluctuations.

Figure 37.1 shows the basic principles of the system applied for automatic tracking of welding joints. Figure 37.1(a) shows the configuration of the system with the bidirectional sensor ($y$ axis, $\Delta y$; $z$ axis, $\Delta z$) connected to the automatic control function which forms position control by driving the $y$ axis and $z$ axis according to the detected values of $\Delta y$ and $\Delta z$. Figure 37.1(b) shows an example of the control signals generated by the system.

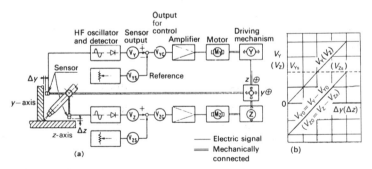

**Fig. 37.1** Basic elements of the electromagnetic sensing system for automatic tracking of welding joints.

### 37.2.1   Heat- and environment-resistant miniature sensor tip

Figure 37.2 shows the heat-resistant sensor (Type C-3). It is constructed with a coreless single-solenoid detecting coil incorporated inside a tightly sealed ceramic structure. Technical features are as follows.

#### (a)   *Heat-resistant, fume- and spatter-proofness*

The tightly sealed ceramic structure means that this sensor tip will have semi-permanent service life even in hot contaminated atmospheres. All the dimensions will be maintained accurately, even in the large temperature fluctuations in the vicinity of the welding arc. The construction also serves to prevent deterioration of the electrical characteristics of the sensor coil.

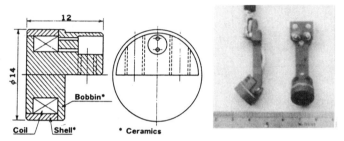

**Fig. 37.2** Heat-resistant sensor (Type C-3).

#### (b)   *Heat and magnetic flux-resistance capability and miniaturization*

Application of the coreless single-solenoid detecting coil makes various technical developments for welding feasible; linearization of characteristics subject to temperature fluctuation; exclusion from the influence of the direct current magnetic field arising from the arc welding current; and linearization of characteristics variation subject to the influence of the alternating magnetic field. Miniaturization of the sensor tip and an extended detection range had already been achieved. All these developments continued to provide an environment-resistant sensing system with high accuracy, supported by the utilization of the compensation function by the control circuits.

### 37.2.2   Temperature compensation

Table 37.1 shows the variation characteristics for both displacement and temperature fluctuation of the electromagnetic sensor coil. The capacitance of the coil is the most affected by changes in temperature, and so this is the main factor in causing temperature drift.

Figure 37.3 illustrates the operational principle of the temperature compensation. This method is designed to superimpose the distance-sensing

**Table 37.1** Variation characteristics of electromagnetic sensor coil

| Equivalent constant | Dependence on distance | Dependence on temperature |
|---|---|---|
| Inductance of coil, $L$ | Large | Null (approx) |
| Distributed capacitance of coil, $C$ | Null | Large |
| Resistance of coil, $R$ | Null | Medium to small |

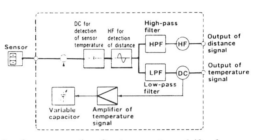

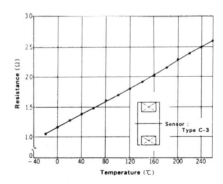

**Fig. 37.3** Principle of compensation for temperature drift of sensor signal.

high-frequency wave (HF) on the temperature-detecting direct current (dc) of the sensor coil. Discrimination of the distance signal output (HF component) from the temperature signal output (DC component) is discussed later. The temperature signal output is then amplified to the required level for input to the variable capacitor, which is designed to compensate the drift of distributed capacitance arising from temperature fluctuations of the sensor coil. Figure 37.4 shows an example of the temperature dependence of the d.c. resistance of the sensor coil. A very small d.c. current will be superimposed, and so the voltage drop thus arising can

**Fig. 37.4** Temperature-dependency characteristics of resistance of sensor coil.

be utilized as the temperature signal output. Figure 37.5 illustrates the correlation between the imposed reverse-polarity voltage and capacitance of vari-cap diode in terms of variable capacitance elements, for which a motor driven variable air condenser equipped with the angle detecting function will also be utilized. Figure 37.6 shows the result of measured values for the temperature compensation; the temperature drift has nearly settled within the range $-30$ to $+500°C$, with the tolerance being within $\pm2\%$ for the target value.

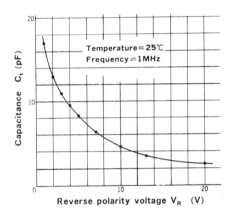

**Fig. 37.5** Correlation between applied reverse-polarity voltage and capacitance of vari-cap diode.

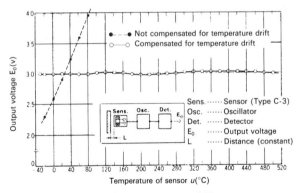

**Fig. 37.6** Result of compensation for temperature drift.

From the above test results, the following features of this temperature compensation can be identified.

1. Because the sensor coil itself is used to detect the temperature, the thermal region for detection is greatly extended.
2. Because a single sensor coil is designed to detect both distance and temperature simultaneously, this allows both the occurrence of

temperature drift for distance detection and variation of d.c. resistance for temperature detection to proceed at the same time. Therefore there is no time lag (or advance) in the compensation function.
3. The sensor tip can be miniaturized, since it is free from any extra attachment except for a single coil.

### 37.2.3  Linearization of the measured output

In terms of output of the electromagnetic displacement meter, measured output usually varies exponentially. A linearizer has therefore been employed for correction.

Figure 37.7 illustrates the operational principle of the linearization method developed here. The measured output is designed to be linearized beforehand by incorporating a simple linearization element in the high-frequency wave circuit, so the system itself is free from the need to have a linearizing device. Figure 37.8 illustrates the characteristics of the bipolar *npn* high-frequency wave transistor in a form of the above linearization. The dotted curves show the conductance between the collector and the emitter, $G_{CE}$, which will be connected to the distance-detecting high-frequency wave circuit for linearizing the measured output. $G_{CE}$ will be adjusted by controlling the base current $I_B$. This effect can be obtained to an approximately equivalent extent either by a bipolar *pnp* or by a field effect transistor (FET). Figure 37.9 shows an example of measured values for the result of linearization, showing favourable results of 0–9 mm for detecting range and $\pm 0.1$ mm tolerance in linearity at the tip of the

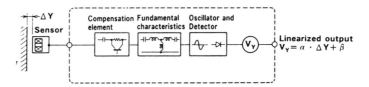

**Fig. 37.7** Principle of linearization of detecting signal.

sensor. From these favourabe test results, the following technical advantages of this linearization method can be identified.

1. The output linearity is good, while the sphere of linearization is extensive. Because the measured output is already linearized by inserting the linearization element in the measuring circuit of the high-frequency wave, including the sensor tip, an extended area can therefore be effectively motivated for output at high linearity.
2. Because the measured output of the high-frequency wave is linearized beforehand, this system can dispense with an extra linearization device.
3. Simply constructed circuits will ensure good performance. This system is composed of a single high-frequency transistor and small parts in restricted number of only two or three.

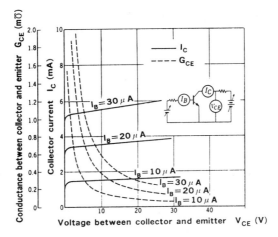

**Fig. 37.8** $G_{CE}$ characteristics of bipolar transistor.

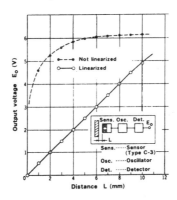

**Fig. 37.9** Result of linearization function.

### 37.2.4 Overall function

Figure 37.10 illustrates the comprehensive block diagram of all the functions described in the preceding sections. As shown in this diagram, all the elements of the sensor tip, temperature compensation and linearization circuits are designed to be connected to the front side of the high-frequency

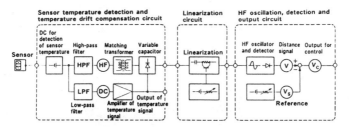

**Fig. 37.10** Comprehensive function block diagram.

circuit, so the output provides linearized signals which correspond directly to the distance.

## 37.3  APPLICATION

A two-channel detecting system is shown in Figure 37.11, used in two of the most common applications: automatic welding apparatus and/or a welding robot for butt and fillet welding joints.

Figures 37.12 and 37.13 show the appearance of a sensor controller and of fillet welding using two-channel sensors, respectively, while Table 37.2 lists the specifications of a two-channel sensor system for general use. The $y$ and $z$ axis sensors are designed to detect the respective distances from the work for correcting the tracking of the welding torch by calculating the deviation either from NC or robot teaching data.

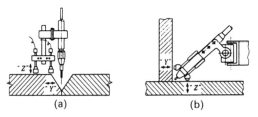

(a)                               (b)

**Fig. 37.11** Two-channel sensing system for general use: (a) application to butt welding joint; (b) application to fillet welding joint.

**Fig. 37.12** Sensor controller for general use.

**Fig. 37.13** Appearance of fillet welding using two-channel sensors.

**Table 37.2** Specification of two-channel sensing system for general use

| | |
|---|---|
| Sensor | Ceramic-shielded single-layer air-core solenoid coil, 14 mm dia × 12 mm (Type C3) |
| Linearization | 0–9 mm, ±1 mm |
| Compensation for temperature drift | −30 to 500°C, ±2% |
| Sensitivity of detection | 0.5 V/mm |
| Reference voltage | 0–5 V, continuously variable |

This two-channel sensor system also can be equipped to detect abnormalities such as overheating or disconnection of the sensor, and to shift the welding torch for fillet welding.

Figure 37.14 shows the sensor of a two-channel sensing system for tracking of a narrow-gap welding joint. This is a complex sensor tip made of two extremely thin sensors with their rear faces bonded together and is used with a differential output controller.

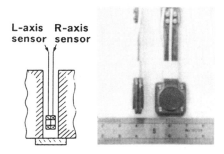

**Fig. 37.14** Sensor of two-channel sensing system for tracking of narrow-gap welding (Type C-4P).

## 37.4 CONCLUSION

We have successfully developed for practical application heat-resistant, environment-proof miniature electromagnetic sensors capable of being mounted on machines in the close vicinity of the welding arc, together with the necessary control apparatus. The principal themes were as follows:

1. development of a heat-resistant, environment- and magnetic field-proof miniature sensor tip;
2. development of a highly accurate compensation method for temperature drift at the sensor tip;
3. development of a measured output linearization method for expanded range;

4. development of a rear face spliced extremely thin two-channel compound sensor tip and of a differential output controller.

However, in order to meet increasingly severe demands on the system, we will continue our efforts toward further miniaturization of the system for increased efficiency.

## REFERENCES

1. Mitsubishi Electric Corporation (1982) *Semiconductors Handbook.*
2. Nippon Electric Corporation (1982) *Electronic Devices Data Book.*
3. Wakamatsu, K., *et al.* (1984) Development of ultra heat-resistant electro-magnetic sensing device and its application to welding apparatuses. *Mitsubishi Juko Giho,* **21**(5), pp. 48–54.

# CHAPTER 38

# Motion generation in an off-line programming system for an arc-welding robot

*Hitoshi Maekawa*

This chapter describes an algorithm for calculating collision-avoiding motion paths for an arc-welding robot. Firstly, safe torch motion is generated independently from the robot in use. A parametric space is introduced to represent torch position on a weld line, and the safety of the corresponding robot position. Then the robot motion, in which the orientation of the weld torch is maintained, is generated by searching in the parametric space. The algorithm is implemented on a personal computer as a tool of an off-line programming system for an arc-welding robot. The algorithm is verified by motion simulation on an object to be welded which has obstacles near a weld line.

## 38.1 INTRODUCTION

In contrast to conventional teaching of a robot by a human operator using real objects, motion generation by a computer using design information for an object and a robot model is called *off-line programming*. The requirements for an off-line programming system are not limited to simple replacement of conventional teaching. Among them are efficient application of CAD data to manufacturing, preliminary productivity checks by appropriate motion simulation, and automatic object code generation for the target robot without interrupting the working production line.

This chapter discusses the methodology for motion generation and some related matters.

*Sensors and Control Systems in Arc Welding.* Edited by Hirokazu Nomura.
Published in 1994 by Chapman & Hall, London. ISBN 0 412 47490 5

## 38.2  SYSTEM CONFIGURATION

By considering the relationships between an object to be welded, the manufacturing tools and a robot, the author has proposed an off-line programming system, which comprises four process layers as shown in Figure 38.1: task level, tool level, robot level and machine-code level [1]. In a typical robot system for gas-metal arc welding, the layers can be described as follows:

1. task level – description of the weld task and its information processing;
2. tool level – motion generation of the welding torch;
3. robot level – realization of torch motion by the target robot;
4. machine-code level – executable code generation for the robot controller.

Motion generation is performed in 2 and 3: tool level and robot level. In the uppermost task level, the welding task is described by a simple programming language. Description of the welding task is important to specify the relationships with other manufacturing processes, but detailed robot information is not required here. At the lowest machine-code level, motion data and welding conditions from the upper two levels are to be compiled into executable code for the target robot.

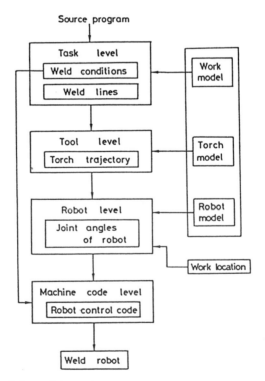

**Fig. 38.1** General flow of the system.

## 38.3  TORCH MOTION GENERATION

At the tool level, the welding torch is regarded as a tool as if it is independent from a specific robot, and its motion is generated to weld to specified lines on an object. The information provided by a welding program, or by interaction between the system and an operator, are the lines to be welded, the starting point and torch position for each line.

Figure 38.2 shows three patterns of obstacle between the torch and an object around a weld line: (a) projection, (b) corner and (c) bridge. These types of obstacle are detected by calculating intersection between the *check face* [2] defined by the welding line and the torch, and each face of the geometric model of the object (the *work model*). Obstacle avoidance in these cases is as follows: (a) projection obstacle and (b) corner obstacle, modification of torch position near the obstacle; (c) bridge obstacle, splitting of the weld line beneath the obstacle. The motion path between weld lines is generated using the check face.

Figure 38.3 shows an example of motion simulation, in which three weld lines are specified. The longest weld line, which runs beneath a *bridge*, is split

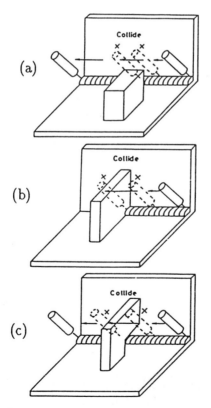

**Fig. 38.2** Some obstacle patterns: (a) projection; (b) corner; (c) bridge.

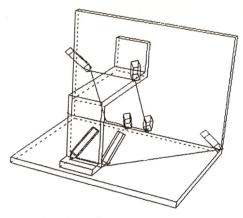

**Fig. 38.3** A simulation of torch motion.

into two segments, and a motion path is generated to step over the bridge. Undesirable changes in torch position and motion path can be simulated at tool level, and this is useful for ensuring adequate design of the joints and welding sequence.

## 38.4  ROBOT MOTION GENERATION

At the robot level, the robot motion is generated to realize the required torch motion for a given location of the workpiece. Once a weld point and a torch position on the point have been determined, the position of the end point $p$ of the robot arm at which the torch-holder is attached is restricted to a certain area in real space. A torch cone is introduced to represent this restriction, describing the relationship between the given torch position and the corresponding arm position [4]. The axis of the cone is coincident with the torch, which is identical to the axis of the torch-head, and its base is the locus of origin of the torch-holder which is attached at $p$: the origin $O_n$ of the $n$th axis of a robot in use. The torch-cone is useful for calculating the position of an articulated robot with five or six degrees of freedom (5- or 6-DOF) to realize the given torch position at a weld point.

For a robot with 5-DOF, it is easy to calculate its pose by regarding the origin of the fifth axis $O_5$ as $p$. But in this type of robot, it is often kinematically impossible to take the required position because of the lack of DOF: in such a case, modification of the location of the workpiece or of the torch position is required [4]. However, for a robot with 6-DOF, the robot position required to realize a given torch position is not determined uniquely while the location of the workpiece is in an appropriate area. Figure 38.4 shows this situation as an example: two different robot positions for a given torch position. In this case, of course, position (b) should be generated.

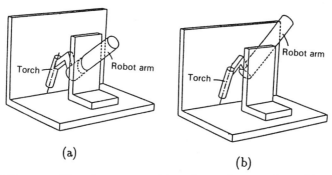

**Fig. 38.4** Two possible robot arm positions for a given torch position: (a) with collision; (b) without collision.

Figure 38.5 shows the structure of a robot with 6-DOF; this type of robot is commonly applied for arc welding. Origins $O_6$ and $O_5$ are aligned on a straight line, and therefore the point $p$ can be taken at $O_5$. Now the problem reduces to determining appropriate $O_5$s, because to calculate the position of a robot with 5-DOF is easier. Figure 38.6 shows a torch cone on a weld line to define the rotation angle of the torch, $\omega$, and the location $x$ of the torch cone along the weld line from the start point. Using these variables, the available $O_5$ is represented as a point in an $x$–$\omega$ parametric space.

An algorithm to determine a sequence for $O_5$ is as follows. First, calculate the availability of robot positions on each digitized node on $x$–$\omega$ space. A

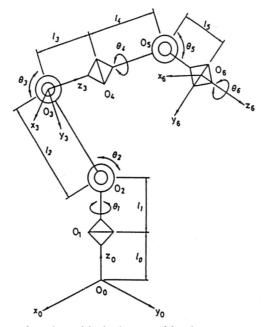

**Fig. 38.5** Structure of a robot with six degrees of freedom.

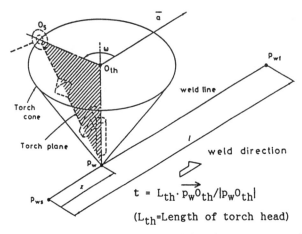

**Fig. 38.6** Rotation angle $\omega$ and location $x$ to define the $\omega$–$x$ parametric space.

node is referred to as *possible* when the corresponding robot pose is kinematically possible, and otherwise as *impossible*. Next, the intersections between the robot arms and the faces of the work model are calculated for each *possible* node, and one that is collision-free is referred to as an *available* node, and otherwise as a *collision* node. Finally, our problem reduces to path searching of the *available* nodes in $x$–$\omega$ space as shown in Figure 38.7.

Paths from the start to the finish point of the weld line are searched by connecting neighbouring *available* nodes in $x$–$\omega$ space from left to right. At this stage, cost functions (penalty weightings, inversely proportional to safety

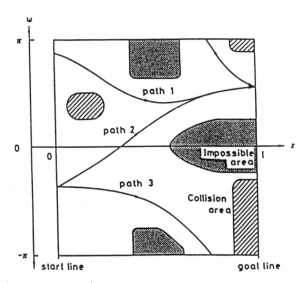

**Fig. 38.7** The $\omega$–$x$ parametric space.

and smoothness of robot motions) of the path are defined as follows by considering the safety and smoothiness of robot motion. The cost for degree of inadequacy of robot position at a node $i$ is

$$d_i = r_1 \times (4 - n_a) \qquad (r_1 > 0) \qquad (38.1)$$

where $n_a$ is the number of available nodes in four neighbours. The cost for a change of robot position from node $i-1$ to $i$ is

$$h_{i-1,i} = r_2 \times \sqrt{(dx^2 + d\omega^2)} \qquad (r_2 > 0) \qquad (38.2)$$

where $dx$ and $dw$ are displacements in the $x$ and $\omega$ directions, respectively. The cost $C_{i,j}$ of the path from $i$ to $j$ is defined as follows:

$$C_{i,j} = \sum_{k=i}^{j} c_k \qquad (0 < i < j) \qquad (38.3)$$

where $c_k$ is the cost of node $k$, given by

$$c_k = d_k + h_{k-1,k} \qquad (k > 0)$$
$$c_0 = d_0 (k = 0) \qquad (38.4)$$

The cost-optimum collision-free path is determined by using these definitions.

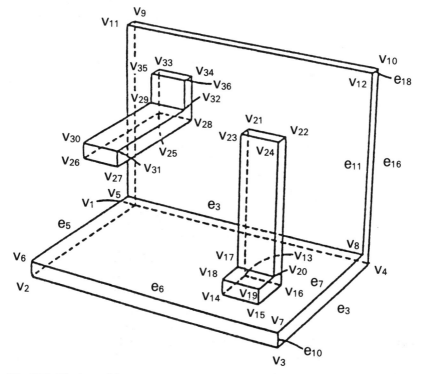

**Fig. 38.8** Work model.

The algorithm is implemented in C on a 16-bit personal computer with an 80286 CPU (10 MHz), 640 kbytes of main memory and an 80287 floating-point coprocessor. Figure 38.8 shows one of the work models used. Figure 38.9 shows the result of an availability and collision check for $8 \times 8$ nodes of an $x - \omega$ space on the weld line $e_8$. There are no *impossible* nodes. *Collision* nodes are represented by triangles, and *available* ones are represented by circles. Robot motion is displayed as animations of skeleton or stereo pairs of robot arms. Figure 38.10 shows a shot of simulation of robot motion. Availability and collision check of the robot position took most of the computation time, which was about 5–6 s for an $8 \times 8$ division.

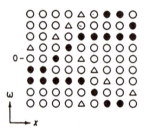

**Fig. 38.9** Result of path search in $x$–$\omega$ space.

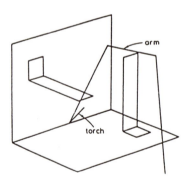

**Fig. 38.10** A skeleton image.

### 38.5  CONCLUSION

Collision-avoiding robot motion paths, in which the torch orientation is maintained at that required for welding, are calculated by the algorithm on a sample object with obstacles near the weld line. The validity of the results has been corroborated by animation of skeleton models of the robot and stereo pair images.

# REFERENCES

1. Maekawa, H., Makino, Y., Kiga, T., Iijima, T. and Nakata, S., (1988) *Off-line programming of arc welding robot in computer aided welding system. Journal of the Robot Society of Japan*, **6**(3), 193–204 (*in Japanese*).
2. Maekawa, H., Yamamoto, K., Kiga, T. and Nakata, S. (1989) *Torch trajectory generation along weld line in off-line programming system of arc welding robot. Journal of the Japan Welding Society*, **7**(2), 102–275 (*in Japanese*).
3. Maekawa, H., Toizumi, A., Kiga, T. and Nakata. S. (1989) *Collision free robot motion generation on a weld line under orientation constraints of weld torch. Proc. 28th SICE*, pp. 1061–64.
4. Maekawa, H., Makino, Y., Kiga, T., Iijima, T. and Nakata, S. (1988) *Motion path processing in off-line programming system of arc welding robot. Journal of the Japan Welding Society*, **6**(2), 245–51 (*in Japanese*).

# CHAPTER 39

# Application of an off-line teaching system for arc-welding robots

*Teruyoshi Sekino and Tsudoi Murakami*

Off-line teaching systems have been newly developed for arc welding robots and applied to robotic welding processes in industrial vehicle and truck assembly lines. Use of the systems reduced the time necessary to teach robots by one-half to two-thirds compared with conventional direct teaching methods. For welding 600 different types of workpiece, a welding FMS was constructed and included a main computer, three robots, and a workpiece carrier. It was demonstrated that off-line teaching systems contributed greatly to system automation and labour reduction.

## 39.1  INTRODUCTION

At present, industrial robots are widely used in many manufacturing sectors and contribute to the improvement of productivity and the reduction of manpower. By operating a robot to work on an industrial object (an actual workpiece) by direct or remote control it is possible to teach the robot about the welding area, position or work sequence. This teaching method has been reliably used because operators can visually check the movement of the robot and the condition of the workpiece during the teaching process. However, in these days of 'flexible manufacturing' (producing a variety of products in small quantities), this teaching method poses some problems, such as the increase in teaching time, the reduction in production efficiency resulting from line stoppage during the teaching work, and danger to the operators.

*Sensors and Control Systems in Arc Welding.* Edited by Hirokazu Nomura.
Published in 1994 by Chapman & Hall, London. ISBN 0 412 47490 5

These problems arise because the teaching must be done by using actual robots or workpieces. As a result, there has been a growing need for an off-line teaching system in which teaching can be carried out by using different devices, rather than by using robots or actual workpieces in the production line. In order to answer these needs, Kobe Steel has developed an off-line teaching system that uses an inexpensive personal computer [1] and has applied this system to scores of actual welding lines. This chapter discusses some examples of its application.

## 39.2 SYSTEM OUTLINE

### 39.2.1 Hardware configuration

The off-line teaching system consists of a personal computer, a CRT display unit, a mouse and a hard disk.

Communication between the personal computer, the robot control panel and other computers is carried through serial lines. Also, the teaching data can be sent and received by the personal computer and the robot through a floppy disk.

### 39.2.2 System operating procedure

The system operation is outlined below. Figure 39.1 shows the operation state and Figure 39.2 provides a simulation example of the robot's motion.

1. *Parts model input*: disassemble the workpiece to be welded into its simple components and input these into the system as parts models.

**Fig. 39.1** The system in use.

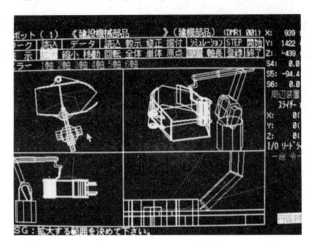

**Fig. 39.2** Simulation of robot motion.

2. *Workpiece synthesis*: assemble the parts models to build a three-dimensional workpiece model.
3. *Operational position teaching*: input a robot's operating loci and the welding conditions applicable to the workpiece model displayed on the CRT to prepare a teaching program.
4. *Robot motion simulation*: display the robot's model and the workpiece model on the CRT to simulate the robot's motion, which has previously been taught, through animations. In this way, check the robot's location, welding position and welding condition. Modify the teaching program if any nonconformance is discovered.
5. *Data transmission*: transfer the prepared teaching program to the robot through the serial communication line.
6. *Execution of the welding*: commence welding with the actual welding robot.

### 39.3  SYSTEM FEATURES

#### (a)  Simple operation

This selection operation system provides ample menus and allows for operation using a mouse. Operators who are not really familiar with personal computers can handle this system easily.

#### (b)  CAD feature

Inputting two-dimensional data in a top view, front view or side view will produce a three-dimensional model when depth data is added. All graphic representations are displayed through surface models after the hidden lines have been processed. Input of data can be done with either a mouse or with numerical data. Straight lines, circular arcs, circles, curves, plate thicknesses, columns, cones and chamfering data may all be put into the system. The input data can be simply converted to other graphic data by using the scaling, rotation, movement or synthesis features.

### (c)   *Program shift*

The teaching program can be moved to an optional location for synthesis. If sliders are attached, data on these sliders can also be moved together. If various workpieces have weld lines that are the same, as shown in Figure 39.3, a programmer need only teach the locus of workpiece 1 and input the amount of the shift for workpieces 2 and 3. Moreover, a programmer can synthesize these programs. Partial scaling is also possible to allow for application to similar workpieces.

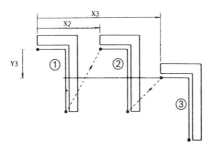

**Fig. 39.3** Program shift.

### (d)   *Mirror image shift*

This feature converts teaching program 1 to programs 2 and 3, which are symmetrical with one another, as shown in Figure 39.4. The robots' wrist angles and slider positions are also simultaneously converted. Usually, when a robot is used to work on workpieces which have weld lines that are symmetrical with one another, it is necessary to teach the robot the path for

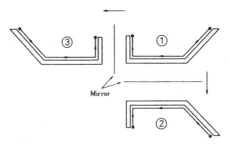

**Fig. 39.4** Mirror image shift.

each component. However, by using the mirror image shifting feature it is necessary to teach only one path, thus improving efficiency.

### (e)   *Robot installation data setting*

This function allows a programmer to adjust the robots and workpieces or peripheral devices to the actual system. The teaching program on the workpiece coordinate system is converted to robot data by using installation data and is then transferred to each robot. Thus a teaching program can be stored for each workpiece, rather than for the robot.

### (f)   *Data transmission*

This is an important function in using the off-line teaching system in factory automation or in a FMS (flexible manufacturing system). The off-line teaching system can be connected to a number of robots and external computers through serial communication lines. This system allows teaching programs to be sent and received by the system and robots. If the system, external computers and robots are connected to one another and operation takes place through a fully automatic transmission mode, programs can be transferred to designated robots on command from external computers, thereby allowing for the possibility of constructing a welding FMS.

### (g)   *Welding condition data*

The automatic welding condition or data bank setting feature allows for the selection of optimum welding conditions. Furthermore, since various sensing functions can be freely selected and set with the system, stable welding quality can be achieved for workpieces which don't always remain steady.

### (h)   *File management*

Much data prepared with the system is stored in hierarchical files. Data management becomes easy and data can be batch-controlled.

### (i)   *Operating environment setting*

The system can establish an environment that is the same as the actual operating environment in every way including the registration of standard/revese setting robots, sliders and positioners. Additionally, as described below, a number of robots can be optionally arranged for one workpiece.

## 39.4 APPLICATION EXAMPLES

The following discusses examples in which the off-line teaching system was applied to welding truck members.

The shapes of the workpieces were of two kinds, represented by A and B in Figures 39.5 and 39.6. Workpiece A includes more than ten different types of part in various sizes and at different mounting locations to be attached to the bottom plate. The welding positions are, therefore, also different. Workpiece B includes more than twenty different types of part, with mounting locations and welding positions that are different for each workpiece which is of a different length.

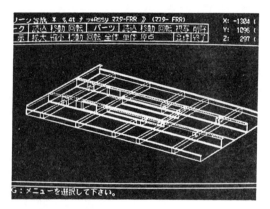

**Fig. 39.5** Workpiece A.

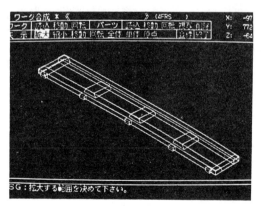

**Fig. 39.6** Workpiece B.

The off-line teaching system is most appropriate for such workpieces. With the system, we can prepare one standard workpiece graphic and a teaching program for each workpiece, A and B, and then, based on the data obtained from this first step, we can easily prepare workpiece graphics and teaching programs for other workpieces by using features such as graphics move, size change, program shift and program synthesis, Figure 39.7 plots the diagram of the off-line teaching system.

Since the depth and width of workpiece A are substantial, the robot operating range required for welding exceeds the range of a single robot. Therefore, robots 1 and 2, placed on sliders, are installed facing each other with the workpiece between them to cover the operating range. The weld line shared by robot 1 and robot 2 is symmetrical to the $x$–$z$ plane. At the same time, for workpiece B, the two robots are arranged on the $x$-axis direction of the workpiece since the workpiece is wider. The weld line shared by robots 3 and 4 is symmetrical to the $y$–$z$ plane.

When preparing teaching programs for these weld lines that are symmetrical to the coordinate planes, it is only necessary to teach one plane. Thereafter, the teaching program for the other plane can be prepared by using the mirror image shift feature as mentioned earlier. Thus, in such cases, only one robot need be taught, thereby taking up only half of the workload compared with the direct teaching method.

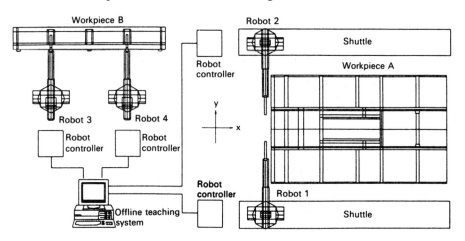

**Fig. 39.7** System application diagram.

## 39.5   APPLICATION RESULTS

The off-line teaching system has been found to be effective as follows after it was used in actual robot welding.

### 39.5.1   Reduction of teaching time

The total teaching time was drastically reduced to one-half or one-third of the time required by the conventional direct teaching method. This is because:

1. The workpieces used were similar in shape, allowing the teaching programs for standard workpieces to be easily converted to the programs

for other workpieces by using the various editing features.
2. The off-line teaching system can correct the installation errors of each robot. Thus, when more than one robot works on an identical workpiece, teaching one robot is sufficient.
3. The example in section 39.4 uses four robots. However, the use of the mirror image feature means that only two robots need be taught.

All of these are the results of the use of the data conversion function, the strongest point of the off-line teaching system.

### 39.5.2   The welding FMS and early line operation

The connection of the off-line teaching system, a number of robots, a production control computer and a workpiece carrier through communication lines makes a fully automatic welding FMS possible. Moreover, off-line teaching work, together with the installation of robots on manufacturing lines, has enabled the operation of welding lines immediately after the completion of robot installation.

### 39.5.3   Improvement of the data management/maintenance function

A large number of programs and welding condition data can be stored in a hierarchical structure on a hard disk, with a large capacity storage, and displayed when required on the CRT. The data management and maintenance therefore become easier.

### 39.5.4   Improvement in the safety of operators

Since most of the teaching work is done off-line, operators need not enter the operating range of robots as much as they used to; therefore, their safety is improved.

## REFERENCES

1. Sekino, T. (1988) *Welding Technology*, **34**(12) 47–50.

# CHAPTER 40

# Development and availability of an IC card welding system for automatic welding

*Kazuhiro Takenaka, Yoshio Imajima*
*and Susumu Ito*

## 40.1  INTRODUCTION

The worsening of the labour market of late has made it especially difficult to obtain skilled workers for the adverse environment of welding. Automation offers a solution to this problem, but in batch production the present situation of welding automation, because of the great variety of products involved, is still dependent on the skill of welding technicians. Therefore, automation tends to be limited to high-volume production of one or a few components. To improve productivity, it is necessary to employ fewer, more highly skilled people, and this means clarifying the concepts of rational and economical welding. With this philosophy in mind, this chapter introduces a welding process control unit and an automatic control unit that employ IC cards as the external memory medium, which have been developed for practical use.

## 40.2  OUTLINE OF SYSTEM FUNCTION

The outline of the system is shown in Figure 40.1.

### 40.2.1  Welding work control unit

The special functions of this work control unit consist of items shown in the specification as items capable of measurement regardless of the welding

*Sensors and Control Systems in Arc Welding.* Edited by Hirokazu Nomura.
Published in 1994 by Chapman & Hall, London. ISBN 0 412 47490 5

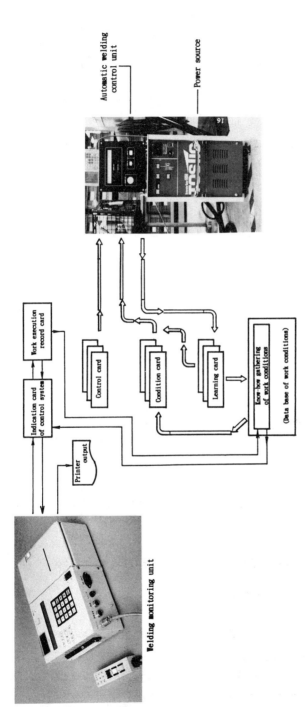

**Fig. 40.1** General configuration of IC card system.

process being applied. These are sampled at intervals of 1/1000 s, and are composed so as to record successively the average value of each point at the distances meeting the purpose for the covering pass and layers.

The control unit is equipped with a relay contact and a device which gives a warning when the work conditions exceed the preset control range during the measurement.

The main specification of the welding work control unit is shown in Table 40.1, and a sample printout of the measured results in Figure 40.2.

**Table 40.1** Main specification

| | | |
|---|---|---|
| Measuring items | | |
| Electrode current | 0–999 A | |
| Electrode voltage | 0–99 V | |
| Electrode feed | 0–9999 mm/min | |
| Wire feed speed | 0–99999 mm/min | |
| Base metal temperature | Max 999°C | |
| | Max 9999°C | |
| Gas quantity detection | | |
| Operation items | | |
| Welding current | Min/max, mean value | |
| Welding voltage | Min/max, mean value | |
| Welding speed | Min/max, mean value | |
| Welding heat input | Min/max, mean value | |
| Wire feed speed | Min/max, mean value | |
| Base metal temperature | Min/max, mean value | |
| | Min/max, mean value | |
| Arc time | Max | Measured for each pass |
| Idle time | Max | Measured for each job |
| Output | | |
| IC memory card | 256 kbit | |
| Printout | 42 | Option |
| Printout paper width | 76 mm | Option |
| Printout relay | 6 ch | |
| Display | | |
| Liquid crystal panel | 40 × 2 | |
| Light-emitting diode | | |
| Alarm | | |
| Communication external dimensions | | |
| Remote control terminal | | |
| Basic unit | 250 × 200 × 120 mm | |
| Printer | 250 × 150 × 120 mm | Option |
| Remocon (excluding card) | 170 × 50 × 30 mm | |
| Power supply | | |
| Basic unit | AC 100 V, 50/60 Hz | |

| Part name | Shield pipe | Wire | Kind SMW308 |
|---|---|---|---|
| Work name | Fitting seat | | Dia 1.2 mm |
| Work date | 1989.06.20 | Electrode for dia | 3.2 mm |
| Worker | Yoshio Imajima | Gas nozzle | Nozzle No.8 |
| Welding length | 200 mm | Ar Gas liquid volume | Nozzle side 10 l/min |
| Measured length | 200 mm | | Back up side 2 l/min |
| Measuring interval | 3 mm | After nozzle | |
| Number of Measured points | | | Ar Gas liquid vol. 1/min |
| Pulse | Setpoint (195) (100) | Electrode pro-trusion length | 15 mm |
| Specification of base metal | Quality of basemetal SUS304 | Distance betw-een base metal and electrode | 3 mm |
| | Plate thickness of base metal 25 mm | Torch angle | Right angle |
| | Quality of part SUS304 | Arc time | 00-19-05 |
| | Plate thickness of part 12 mm | | |

| Pass No. | Point No. | Welding current(Amp) | | Welding voltage | Welding speed mm/min | Ware quantily cm/min | Welding input heat value J/cm | Temp of base metal °C | Temp of molt-in p-ool C |
|---|---|---|---|---|---|---|---|---|---|
| | | Peak current | Base current | | | | | | |

**Fig. 40.2** Sample printout from the system.

### 40.2.2 Automatic welding control unit

The characteristic function of this unit consists of IC cards whose composition corresponds to the universal control unit and the driving part of each automatic machine. There is also a conditions IC card that corresponds to the shape, material plate thickness, etc. of weldments, and a learning IC card that can add corrections specific to the job site to the basic conditions card.

Maintenance can pose problems for automatic machines as well; IC cards for maintenance are provided so that even a novice can perform maintenance.

An application example of the automatic control system of this unit is shown in Figure 40.3.

### 40.2.3 Power supply for automatic welding unit

The power supply to this welding unit is an a.c./d.c. welding power double inverter system, and all its control factors can be supplied directly from the condition cards. Therefore, even an inexperienced worker can set the power supply, setting unerringly, and achieve excellent repeatability of conditions. Considerations are given so that the welding conditions can be produced even for the first welding by a rational method concentrating on the welding quality.

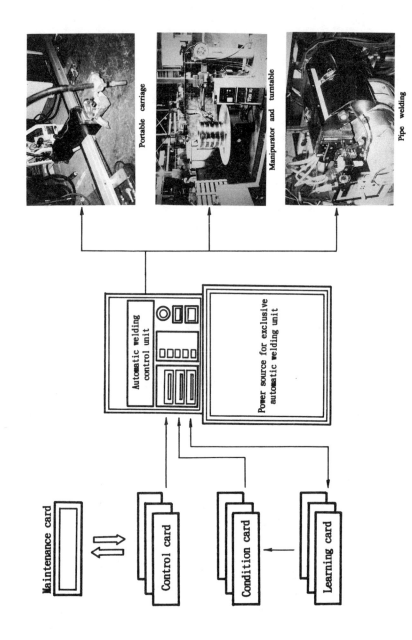

**Fig. 40.3** Typical practical application of the control system.

## 40.3   COMPOSITION AND FUNCTIONS OF EQUIPMENT

### 40.3.1   Welding work control unit

The internal structure and outline connection diagrams of the welding work control unit are shown in Figures 40.4 and 40.5.

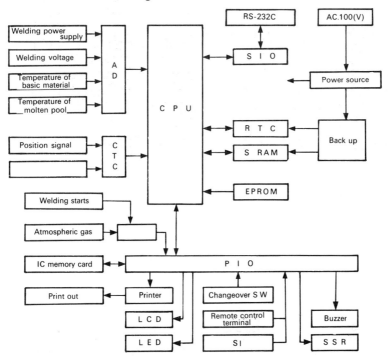

**Fig. 40.4** Block diagram of control unit.

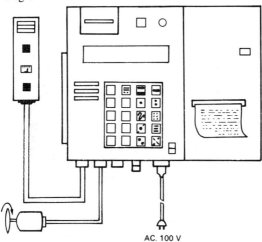

AC. 100 V

**Fig. 40.5** Sketch of the connections to the control unit.

Of the measuring factors, the welding current ($I_p$, $I_b$), welding voltage, temperature 1 (low temperature), and temperature 2 (high temperature) are represented by analogue signals; the welding speed and wire feed speed, by digital signals; and the detection of the presence or absence of shield gas, by on/off signals.

There are two methods of input to the system. The control range and order or execution can be input by IC cards; or the work order can be chosen freely using ten keys.

For output, the recorder IC card is read with personal computer software, and the printout is made by arranging all the measuring points, passes and layers or by arranging only the mean values of the passes and layers. Alternatively, the measured results can be printed out in sequence.

Figure 40.6 shows the measuring factors and the relationship between output point, pass and layers.

If the work conditions deviate from the preset control range, an error record is registered in the measurement record and the alarm contact and a buzzer operate.

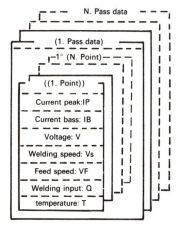

**Fig. 40.6** Measuring factors and relationship between output point, pass and layers.

### 40.3.2  Automatic welding control unit

This control unit differs from conventional automatic welding control in the processing method that is used.

It comprises the condition IC card which selects the control IC card which in turn controls the driving source targeted by the IC-card universal control device and the learning IC cards that form the conditions specified to the job.

Conventional control devices are designed to work with a specific automatic drive mechanism. This unit can be applied to any automatic machines (satisfying the specification of the control device) by exchanging the IC cards.

Its working conditions can be quickly and rationally optimized through interaction between the basic condition IC cards and the learning IC cards.

Maintenance can pose a problem in this kind of control device. Even a novice can detect faults caused by an elementary misoperation through the use of maintenance IC cards.

Figure 40.7 shows the functional configuration of this system.

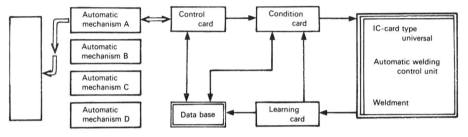

**Fig. 40.7** Configuration of welding control unit.

## 40.4   MAIN FEATURES OF APPLICATION EQUIPMENT

### 40.4.1   Welding work control unit

This unit permits a systematic analysis of working conditions, the scientific accumulation of work data, and rational instructions for the conditions of new weldments.

1. The major factors for welding work can be measured simultaneously by means of a single measuring instrument.
2. Because of the sampling intervals of 1 ms, sufficient analysis data can be obtained.
3. Control items and ranges can be preset, thus permitting detailed work records.
4. Out of range conditions can be detected and a warning alarm given.
5. Detailed record data are processed with a personal computer, thus facilitating quick analysis.

### 40.4.2   Educational training equipment

The training of welding technicians requires a skilled teacher and a long training period. In this training equipment, by registering the optimal training condition in the IC cards, systematic training is available without any time restrictions. Its main features are as follows:

1. Training for practical skill is affected merely by preparing a training card in the work control equipment.
2. The results of training are recorded in the card for subsequent review.

3. An effective educational training requires a systematic accumulation of knowledge and experience and this unit is optimal for the purpose.
4. Even in the absence of a specially trained teacher, education can be conducted through a scientific procedure.
5. The training system can be easily expanded by adding extra equipment.

### 40.4.3  Automatic welding control unit

The use of control IC cards makes it possible to use this unit to control many different applications. Two examples are considered below.

### (a)  *Automatic pipe welding device*

Applications of welding to pipework are numerous and varied, and many different products have been introduced to the market, but only a few are in common use. Most of the types sold on the market are specific to a particular application, and this has restricted their popularity.

However, if this control unit is adopted, by preparing IC cards to match the welding head, the equipment of a dedicated system can be realized. The main features are as follows.

1. It is a universal unit providing good economics of operation.
2. By exchanging the control IC cards, it is easy to change the application, and there is the added benefit of space saving.
3. The working conditions can be processed with a personal computer, and repeatability and quality control can be easily achieved.
4. Through the learning function and personal computer processing, it is easy to deal with new products systematically.
5. Utilization of a maintenance IC card enables quick and easy judgement of the degree of damage, even by the inexperienced.

### (b)  *Automatic welding unit for special shapes*

This unit can, as a matter of course, control a general welding tractor and welding heat. Hence, by placing an elliptical welding subject in the centre of a ring-shaped rail, it is also possible to cope with various shapes. A general view of the mechanism is shown in Figure 40.8. The main features are as follows.

1. By the use of a personal computer, it is possible to fashion condition IC cards freely in the shape of weldments.
2. Even for a new material, rational and optimum conditions can be yielded with the learning IC cards.
3. Weldments of complicated shape and difficult welding materials can be dealt with.

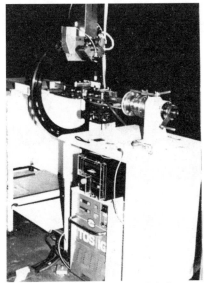

**Fig. 40.8** Application to special shapes.

4. Within the working range of the welding mechanism, various welding work can be done with the same welding device.
5. The work data can be freely processed in the personal computer.

## 40.5 CONCLUSION

It is evident that welding automation will advance at an increasing tempo in response to the conditions of each industrial field. However, it should not be centred on the mechanization of simple work, but should be an automation system adaptable to scientific techniques, independent of the experience and skill of workers.

This chapter has introduced part of the equipment developed by our company for practical use along these lines.

# CHAPTER 41

# Development of a remote-controlled circumferential TIG welding system

*Tomoya Fujimoto and Atsushi Shiga*

A remote-controlled circumferential TIG welding system, monitored through images and sounds of the arc and the surrounding area, has been developed. This system has been applied to the field welding of water supplying SUS pipes and gas pipes. Consequently, the present system has proved successful both in practical application and welding monitoring.

## 41.1  INTRODUCTION

Notable development effort has been put into saving manpower and energy, reducing cost, and increasing the efficiency of welding. This chapter discusses a remote-controlled circumferential TIG welding process, which ensures a fine adjustment of welding parameters during welding in real time through the monitoring of a coloured image and the arc sounds of the molten pool.

## 41.2  MONITORING EQUIPMENT

### 41.2.1  Structure

Figure 41.1 shows the general structure of the system. A carriage on which a welding head is mounted is positioned on a guide ring fixed around the pipe.

*Sensors and Control Systems in Arc Welding.* Edited by Hirokazu Nomura. Published in 1994 by Chapman & Hall, London. ISBN 0 412 47490 5

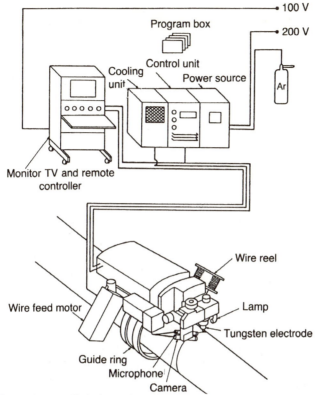

100 V

Program box

200 V

Cooling
unit

Control unit

Power source

Ar

Monitor TV and remote
controller

Wire reel

Wire feed motor

Lamp

Tungsten electrode

Guide ring

Microphone

Camera

**Fig. 41.1** Remote-controlled circumferential TIG welding system.

Away from the trench are located a monitor TV along with a speaker and a remote controller, a welding power source, a control unit, and a torch cooling unit.

During welding, fine adjustments of the positions of filler wire and torch, welding current, arc voltage, wire feed rate and welding speed are performed appropriately by an operator through observing the monitored images and sounds. These operations are manipulated on the controller board.

Figure 41.2 shows a schematic illustration of the present monitoring system. The structure of the welding head is shown in Figure 41.3. A micro camera for taking pictures of molten pool, filters, a lamp, and a microphone is installed around the welding head. The specifications of the monitor unit components are indicated in Table 41.1.

### 41.2.2  Features

The features of the system are as follows:

1. real-time remote-controlled welding system using TV monitor images and sounds;

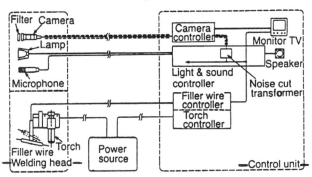

**Fig. 41.2** Schematic illustration of monitor system.

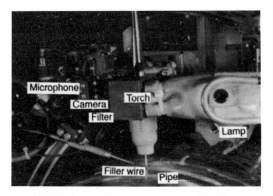

**Fig. 41.3** Structure of welding head.

2. maximum height of welding head only about 200 mm;
3. correct welding condition is set up before welding;
4. Deep penetration [1] secured by pulsed arc welding method;
5. Welding scene is continuously recorded on video disk.

### 41.2.3  Welding practice

#### (a)  Power supply

The power supply conforms to a synchronized-pulsed d.c. arc welder [2]. The output current ranges from 5 to 300 A, and the maximum output voltage is equivalent to 35 V.

#### (b)  First-pass welding

It is essential to obtain a sound internal bead in root pass welding for a fixed pipe. The levels of welding current and arc voltage optimum for each welding position, the condition of molten pool, and the interrelations

**Table 41.1** Specifications of monitor unit components

| | |
|---|---|
| *Camera* | *CCD, about* $3 \times 10^5$ *elements, 25 grf* |
| Filter | Infrared rays cut, band pass and interference |
| Lamp | 100 V, a.c. 150 W, halogen |
| Microphone | Condenser type |

between the positions of the tungsten electrode, the filler wire, and the front line of molten pool, must all be accurately selected. Precise control is essential to prevent welding defects.

### (c) Second- to final-pass welding

Both reasonable penetration and a normal bead appearance are required for each pass: hence fine control of weaving conditions, wire feed rate and heat input is essential. Examples of the monitor image and the internal penetration bead appearance are shown in Figures 41.4 and 41.5 respectively.

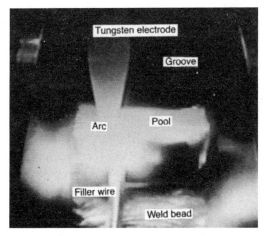

**Fig. 41.4** Example of monitor image.

## 41.3 EFFICIENCY OF MONITORING

The use of the present system, which features monitoring through welding images as well as through sounds, produces the following efficiencies.

1. All-position welding of fixed pipe is feasible by utilizing only basic techniques and knowledges; it does not require great experience or operator skills.

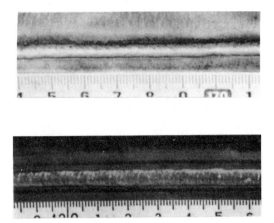

**Fig. 41.5** Examples of internal bead appearance: (a) SUS 304; (b) SGP.

2. Welding defects such as undercuts, overlaps, unfavourable reinforcements, inclined beads, arc blows, and the bending of filler wires, are immediately modified through the monitor images to change the welding conditions.
3. Good welding practice is realized continuously by ensuring optimum welding conditions at all times, preventing the generation of welding defects by means of diagnosing the alteration of welding sounds. These include arc sound, the contact noise of the wire with the weld bead, a derivative sound issued as the tungsten electrode gains access to the filler wire, and the sounds of motors.
4. The conditions of the molten pool during welding can be recorded on video tape for post-welding investigation.
5. The remote-controlled system can be used for circumferential welding in deep, narrow trenches, thus eliminating the potential safety hazards of working in the trench.

## 41.4   RESULTS

### 41.4.1   Welding on water supply pipe

The pipes tested were of stainless steel, grade SUS 304 with 400 mm outside diameter, 8 mm wall thickness and 6000 mm length. The groove geometry is shown in Figure 41.6 and the welding conditions and arc time are shown in Table 41.2. The filler wire was of 1 mm diameter, and the shielding gas for the arc as well as inside the pipe was Ar with a flow rate of 10–15 l/min.

The principal results obtained were as follows.

1. Three welded pipe connections a day were achieved.
2. The properties of the weld joints were equal superior to the Second Grade of JIS (Japan Industrial Standard).

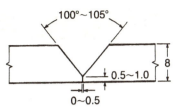

**Fig. 41.6** Groove geometry.

**Table 41.2** Welding conditions and arc time for SUS 304 pipe

| Pass | Welding current (A) | | Arc voltage (V) | | Travelling speed (mm/min) | Dwell time (s) | | | Wire feed rate (mm/min) | | Arc time (min) |
|---|---|---|---|---|---|---|---|---|---|---|---|
| | IP | IB | VP | VB | | TL | TC | TR | WP | WB | |
| 1 | 100–105 | 95–100 | 9.0 | — | 85 | 0 | 0 | 0 | 650 | 750 | 15 |
| 2 | 210–220 | 60–70 | 9.3 | 6.5 | 75 | 0.4 | 0.4 | 0.4 | 1050 | 1000 | 17 |
| 3 | 220–230 | 90–95 | 9.3 | 7.3 | 70 | 0.4 | 0.4 | 0.4 | 1750 | 1300 | 18 |
| 4 | 230–240 | 90–95 | 9.3 | 7.3 | 65 | 0.5 | 0.5 | 0.5 | 2050 | 1550 | 20 |
| 5 | 200–220 | 115–125 | 9.3 | 7.3 | 50 | 0.5 | 0.5 | 0.5 | 1650 | 1050 | 25 |
| | | | | | | | | | | Total | 95 |

3. Stable one-side welding is achieved even with no root opening by holding the groove angle wider than one hundred degrees.
4. The present remote-controlled welding system has been found useful in practice, and enables continuous welding to be achieved.

The remote control operation during welding in the field is shown in Figure 41.7 and the welding head on the travelling unit is shown in Figure 41.8.

**Fig. 41.7** Remote-control operation during welding.

**Fig. 41.8** Automatic girth welder in the trench.

### 41.4.2    Welding on gas supply pipe

Mild steel pipes 7.9 mm in thickness, 350 mm outer diameter with 5000 mm length were used. The groove geometry was the same as in Figure 41.6. The maximum offset was controlled to within 1 mm.

The pass sequence as well as the build-up condition are shown in Figure 41.9. The welding conditions and arc time are shown in Table 41.3. The welding wire used had a diameter of 1 mm.

The results were as follows.

1.  It took 50–60 min arc time for a girth weld.

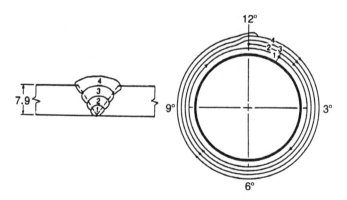

**Fig. 41.9** Pass sequence and build-up conditions.

**Table 41.3** Welding conditions and arc time for SGP pipe

| Pass | Welding current (A) | | Arc voltage (V) | | Travelling speed (mm/min) | Dwell time (s) | | | Wire feed rate (mm/min) | | Arc time (min) |
|---|---|---|---|---|---|---|---|---|---|---|---|
| | IP | IB | VP | VB | | TL | TC | TR | WP | WB | |
| 1 | 145–150 | 140–145 | 8.8 | — | 100 | 0 | 0 | 0 | 700 | 760 | 11 |
| 2 | 280–290 | 110–120 | 10.5 | 8.5 | 95 | 0.4 | 0.4 | 0.4 | 2410 | 1270 | 12 |
| 3 | 290–300 | 110–120 | 10.5 | 8.5 | 85 | 0.5 | 0.5 | 0.5 | 2410 | 1500 | 13 |
| 4 | 280–290 | 120–130 | 10.5 | 8.5 | 80 | 0.5 | 0.5 | 0.5 | 2410 | 1500 | 15 |
| | | | | | | | | | | Total | 50 |

2. The results of radiographic inspection of 65 sheets of film, were that 63 sheets showed superior to Second Grade welding, and two sheets showed some porosity equivalent to the Third Grade in JIS.
3. Welding of pipes inclined at 2°–5° was feasible without any trouble by the remove control operation.

The view during welding is shown in Figure 41.10, and the bead appearance is shown in Figure 41.11.

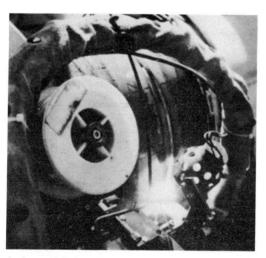

**Fig. 41.10** View during welding.

## 41.5 CONCLUSION

A remote-controlled circumferential TIG welding system has been developed which is monitored through TV images and sounds of the arc and the surrounding area.

**Fig. 41.11** Example of external bead appearance.

An investigation of one-man operation and trial applications to GMAW and FCAW are expected to be made in the future in order to realize a welding system superior in cost saving.

## REFERENCES

1. Yamamoto, Y. *et al.* (1981) On recent semi-automatic $CO_2$ and MAG welding machine. *Welding Technique*, **29** (7), 21–8.
2. Imaizumi, K. (1986) On recent automatic pipe and tube welding system. *Pipe Arrangement and Apparatus*, No. 3, 7–15.

# CHAPTER 42

# In-process control systems in one-side SAW

*Hirokazu Nomura, Yukihiko Satoh*
*and Yoshikazu Satoh*

Nowadays, there is great demand for excellent welding technology which will provide high weld quality and high efficiency, without the need for skilled operators. To meet this requirement, NKK has conducted a great deal of research and development on welding automation. This chapter describes the features of in-process control systems for one-side SAW developed by NKK, such as back bead width controls in TCB-SAW and FCB-SAW, and face bead height control in TCB-SAW.

## 42.1   INTRODUCTION

For welding automation to be automated in the true sense of the word, and not merely mechanized, there are two methods of carrying out automated control.

In one method, the welding parameters and torch position are corrected while the state of the groove is detected (or detected and corrected in advance). For this method, stability and high-accuracy setting of the variables of the welding equipment are essential. It is also essential to clarify how the welding results will change when the welding parameters are varied, and how to change the welding parameters to cope with changes in the state of the groove. A control algorithm-based theorem must be established.

In the other method, while directly or indirectly detecting information on the welding results (for example, bead width or weld penetration), the welding parameters are directly controlled so as to direct the welding results

*Sensors and Control Systems in Arc Welding.* Edited by Hirokazu Nomura.
Published in 1994 by Chapman & Hall, London. ISBN 0 412 47490 5

towards the target quality. With this method, it is possible to carry out control with a simple control algorithm. However, the choice of welding control information, and the means of obtaining this, determine the effectiveness of the control. Consequently, it may well be said that detection of the physical quantities corresponding to the welding result and the means of detection are the keys to the success of control.

The control system described in this chapter has been developed according to control ideas based on the latter way of thinking.

We have developed a control system capable of detecting the physical quantities corresponding to the welding result and feeding back to the principal welding parameters so that, in one-side SAW, uniform and stable back bead and face bead shapes can be obtained despite any fluctuations in groove shape, including root gap, groove angle or the presence of tack-welding beads, or the presence of any external disturbances such as fluctuations in the primary power voltage.

## 42.2 ARC LIGHT INTENSITY CONTROL OF BACK BEAD WIDTH [1]

A tape and copper backing process using glass fibre (TCB process) has recently been employed for the one-side welding of plates with inclined weld lines or poor groove accuracy (for example, welding of curved shell or erection joints). This is a two-electrode welding process in which, with the groove filled with iron powder, penetration and back bead are formed with the leading electrode and the face bead with the trailing electrode so that welding of a plate can be performed with the face and back beads completed by just one pass.

NKK has developed a new backing process (self-driving backing carriage system), an improvement of the conventional TCB system, as shown in Figure 42.1, and has put it to practical use. Conventionally, the backing material and backing strip is fixed over the entire weld line. However, the striking feature of this new backing process is that a small copper backing

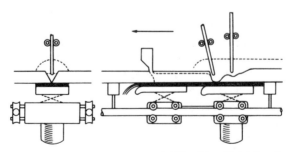

**Fig. 42.1** Schematic illustration of tandem SAW with self-driving backing carriage system.

plate locally supports the formed back bead, and is driven by a motor, travelling synchronously with the welding carriage (with the progress of arc).

The arc light intensity control system has been developed for the TCB process using this self-driving backing carriage system.

### 42.2.1 Principle of back bead width control

In the TCB process, almost no arc light is observed where no back bead has been formed, but arc light is clearly observed on the backing side where a back bead has been formed. From this observation, it can be assumed that under welding conditions in which a satisfactory back bead can be obtained, the back bead is formed mainly by a keyhole action [2]. Furthermore, there is a quantitative correlation between the brightness of this arc light and the back bead width.

Figure 42.2 is a schematic drawing of the arc intensity detecting process. As is evident from this diagram, four photosensors are embedded in the copper backing plate in diagonal directions with the arc generating point as the centre, so that the intensity of the arc light transmitted through the glass-fibre tape by way of the molten pool keyhole is detected.

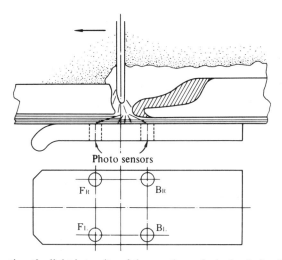

**Fig. 42.2** Detecting the light intensity of the arc through the keyhole of the weld pool and glass-fibre tape.

The relationship between the detected arc light intensity and the back bead width is as shown in Figure 42.3. The abscissa represents the sum total of the output voltage of the four photosensors, or the physical quantity corresponding to the arc light intensity. The said correlation is obtained with the welding current being varied over a wide range for the three types of groove conditions, as shown in Figure 42.3, and good correlation is seen

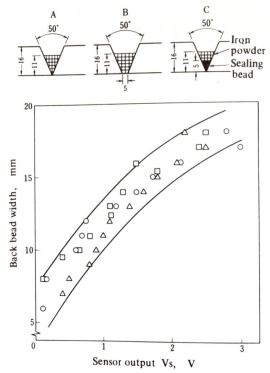

**Fig. 42.3** Correlation between arc light intensity and back bead width: ○, groove A; △, groove B; □, groove C. Welding speed, 310 mm/min; arc voltage, 32 V; plate inclination, 0° horizontal.

between the arc light intensity transmitted through a glass-fibre tape and the back bead width. It is also seen that this correlation is unaffected by the presence of tack welds and root gap size. In other words, it may well be said that the arc light intensity directly represents the welding result of the back bead width.

### 42.2.2   Development of back bead width control system

NKK has developed a back bead width control system based on the relationship shown in Figure 42.3. Figure 42.4 is a block diagram of the control system. The leading welding current is feedback controlled so that the arc light intensity can be kept at a preset level, $V_{ref}$, and other welding parameters are fixed. This control circuit is characterized by including the welding phenomena within the control loop. It may well be said that this control system performs feedback control directly on the arc phenomena so as to maintain a constant back bead width.

Figure 42.5 shows the effect of this back bead width control. The abscissa represents the controlled welding current. As is evident from the diagram,

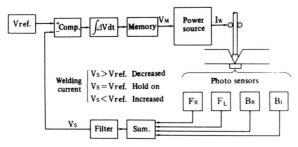

**Fig. 42.4** Block diagram of welding current control based on detection of arc light intensity.

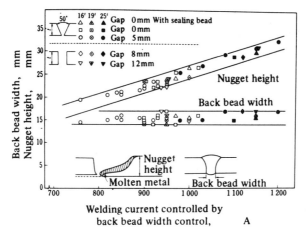

**Fig. 42.5** Effect of back bead width control on welding current.

the welding current changes according to different types of plates with widely varying groove states, with the result that a uniform back bead width can be obtained over a wide range of groove states. Furthermore, it is verified that the back bead width is controlled at a uniform value despite sudden changes during welding.

## 42.3 VOLTAGE DETECTION CONTROL OF BACK BEAD WIDTH [3]

NKK has developed a back bead width control system applicable to the flux and copper backing process (FCB). This is a typical fixed backing one-side welding process which has been adopted at most of the major and medium-sized shipyards in Japan.

As a result of our analysis of the welding phenomenon in an FCB process, it has been established that where stable keyhole welding is performed with the welded material electrically insulated from the backing plate, an electric potential of several volts is detected between the copper backing plate and the baseplate being welded, and that this potential difference correlates with the back bead width. By studying this detected voltage (hereinafter called 'Detecting Voltage or Vd'), we developed a back bead width control system.

### 42.3.1    Principle of back bead width control

When FCB one-side welding is performed under welding conditions in which sufficient back beads can be formed, if the electric voltage is detected by the procedure shown in Figure 42.6(a), a detected voltage ($V_d$) waveform as shown in Figure 42.7 is obtained. An arc voltage, $V_a$, is also apparent in the upper stage, representing the polarity with the baseplate as a common 'cable'. One of the most important characteristics of FCB one-side welding is that the polarity and phase of $V_d$ coincide with those of the arc voltage.

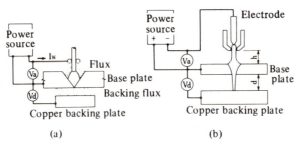

(a)                                (b)

**Fig. 42.6** Measurement of the voltage between baseplate and copper plate: (a) SAW, (b) PAW.

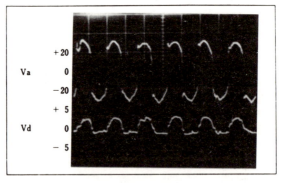

**Fig. 42.7** Typical waveform of $V_d$ and $V_a$.

In order to clarify the cause of generation of $V_d$, in transfer type plasma arc welding (PAW) as shown in Figure 42.6(b), measurements were taken with the baseplate electrically insulated against the copper plate in the same manner as for SAW. As a result, $V_d$ of about 2 V was detected as the plasma transmitted through the keyhole reached the copper plate.

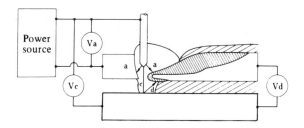

**Fig. 42.8** Arc voltage distribution in one-sided FCB-SAW: $V_a = V_c + V_d$.

From the result described above, a model of the cause of generation of $V_d$, is prepared as shown in Figure 42.8. As is observed in plasma arc welding, a considerable amount of plasma reaches the copper plate, as shown by 'c' in the drawing. It can be assumed, therefore, that $V_d$ is generated as a result of the plasma contacting the baseplate and that it corresponds to the electric potential of a part of the arc column. Of course, the route represented by 'a' in the drawing may be also present.

The relationship between $V_d$ and back bead width is quantitatively obtained as shown in Figure 42.9. $V_d$ is not detected in region 'b', possibly because the back bead is formed in a convection action. In region 'a', a linear relationship between $V_d$ and back bead width is observed. Consequently, as long as keyhole welding is performed, $V_d$ is thought to be significant as a physical quantity representing back bead width.

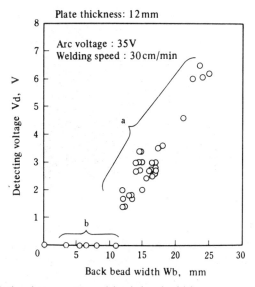

**Fig. 42.9** Correlation between $V_d$ and back bead width.

### 42.3.2 Development of back bead width control system

We have developed a system for controlling welding current so as to make $V_d$ agree with a preset value by making use of the relationship between $V_d$ and back bead width shown in Figure 42.9. A block diagram of this control system is shown in Figure 42.10.

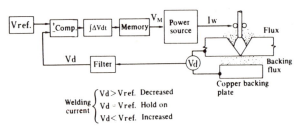

**Fig. 42.10** Block diagram of welding current control based on detection of $V_d$.

As shown in Figure 42.11, three types of groove were prepared on a weld line: a standard V-groove with zero root gap, a V-groove with a 3 mm root gap, and a V-groove with an 8 mm throat depth seal bead. The results of one-side welding with and without back bead control (with a fixed welding current) are shown. It can be seen that where the control is applied, satisfactory results are obtained for the standard V-groove and the seal bead, but for the 3 mm root gap, the feedback is too effective and the back bead width is excessively reduced. This is because the presence of the root gap allows easier penetration of the arc 'flames', with a resultant decrease in the welding current. As a result, the molten pool head becomes smaller, so that it becomes easier for the arc 'flames' to penetrate even with a weak arc force. By combining the face bead height control described in Section 42.4, it may be possible to improve this point.

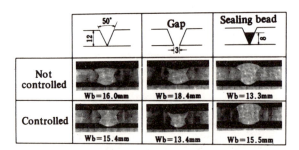

**Fig. 42.11** Effect of welding current control based on detection of $V_d$.

## 42.4 DEVELOPMENT OF FACE BEAD HEIGHT CONTROL SYSTEM [4]

With the back bead width control system described above, in each case the arc force is controlled by changing the welding current so that a constant back bead width can be obtained. In this control system, the welding current automatically changes with fluctuations in the state of the groove. However, this changing of welding current has a shortcoming, in that it aggravates the variation of reinforcement of the weld, because the other welding parameters are fixed while the welding current automatically changes.

We have therefore developed a face bead reinforcement height control process in which this shortcoming is eliminated by feeding a filler wire into the arc of the leading electrode. By combining this process with the back bead control system, we have developed a system capable of complete simultaneous control of the face and back beads.

### 42.4.1 Principle of face bead height control

Figure 42.5 shows the relationship between the controlled welding current and the back bead width and nugget height where the arc light intensity type of back bead width control is performed. The term 'nugget height' used here is defined as the throat thickness from the bottom surface of the back bead to the top surface of the face bead. It can be seen from the diagram that where the back bead width is controlled at a constant value, a linear relationship exists between the welding current and the nugget height, or that it is free of any influence from plate thickness and groove condition. As a result, it may be said that the welding current controlled by the back bead width control represents the physical quantity of the nugget height.

Figure 42.12 is an explanatory drawing of this relationship. In general, it is known that the arc force, which is dependent mainly on the welding current, balances well with the specific gravity head ($\rho gh$) of the molten pool. Where the back bead width is controlled at a constant value and the keyhole is stably formed, as shown in the drawing, the molten pool head height

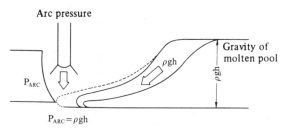

**Fig. 42.12** Balancing behaviour between arc pressure and gravity head of molten pool.

balanced by the arc force coincides with the nugget height. Consequently, there is a good correlation between the welding current and the nugget height.

Incidentally, the nugget height is the sum total of the face bead reinforcement height, the plate thickness and the back bead reinforcement height, and since the plate thickness and back bead reinforcement height can be determined in advance, the controlled welding current is a physical quantity representing face bead height.

It can be seen from Figure 42.5 that, where face bead height is low, if certain measures are provided to increase the controlled current value, the face bead height ought to increase. Figure 42.13 shows the result of such a measure being taken; a 2.4 mm diameter cold wire is fed into the arc of the leading electrode while controlling back bead width. As is evident from the diagram, an increase in the feed rate of the cold wire causes both the welding current and the nugget height to increase, although the back bead width remains constant, showing that addition of cold wire is effective for controlling face bead reinforcement height.

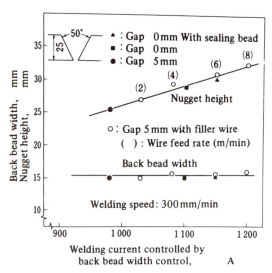

**Fig. 42.13** Effect of addition of filler metal.

### 42.4.2  Development of face bead reinforcement height control system

Figure 42.14 is a block diagram of an arc light intensity control system capable of simultaneously controlling back bead width and face bead height. This system is designed to detect the arc light intensity on the backing side and to control the welding current so that the arc light intensity becomes constant at the specified $V_{ref}$ while at the same time controlling the filler wire

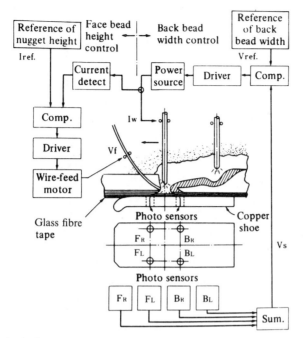

**Fig. 42.14** Block diagram of nugget dimension control system.

feed rate so that the said current coincides with the specified $I_{ref}$. The values corresponding to the required back bead width and nugget height are input as $V_{ref}$ and $I_{ref}$ respectively.

Figure 42.15 shows the effect of simultaneous control of face and back beads. As is evident from the photomicrograph of the cross-section, this control system is free of any influence from the presence of tack welds or root gap fluctuations. The control system can also obtain almost the same weld nugget with an I-groove, thus showing its capability of meeting a wide range of groove conditions. Furthermore, it has been verified that the nugget form can be uniformly controlled if any sudden changes occur during welding.

## 42.5 STATE OF PRACTICAL APPLICATION

The TCB back bead control system has been applied to welding of thick-wall steel pipe piles and orthotropic steel bridges, and the face and back beads simultaneous control system has been applied to the welding of orthotropic steel bridges [5]. FCB back bead control is scheduled to be put to practical use in the near future in the construction of a large assembly parallel panel line at the Tsu Works.

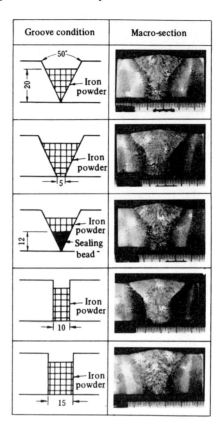

**Fig. 42.15** Effect of nugget dimension control.

## 42.6   CONCLUSION

In the one-side TCB and FCB welding processes, we have developed a control system capable of controlling the back bead at a constant value. In the TCB process, we have developed a face and back beads simultaneous control system, capable of controlling face bead height at a constant value, in combination with the back bead width control.

## REFERENCES

1. Nomura, H. *et al.* (1979) Development of feedback current control system in one-sided submerged arc welding. *Nippon Kokan Technical Report. Overseas* No. 27.
2. Nishiguchi, K. (1970) The mechanism and estimation of back bead formation. JWS-Arc Physics Commission/JIW-212, 19th.

3. Nomura, H. (1985) Back bead width control in one-side SAW using flux copper backing. *Journal of the Japan Welding Society*, **3** (3) 35–41.
4. Nomura, H. *et al.* (1982) An automated control system of weld bead formation in one-side submerged arc welding. JWS-Welding Process Commission, SW-1340-82, 87th.
5. Nomura, H. *et al.* (1985) Automated control system in one-side SAW for orthotropic type bridge. JWS-Welding Process Commission, SW-1653-85, 105th.

# CHAPTER 43

# Application of an arc sensor and visual sensor on arc welding robots

*Teruyoshi Sekino*

This chapter discusses the short arc sensor and the visual sensor used for the high-speed welding of thin plates as application examples of torch positions/loci correction for robot arc welding.

## 43.1 INTRODUCTION

The recent spread of factory automation or FMS for arc welding operations is remarkable. One of the factors causing this spread is the improvement in and practical availability of sensing techniques. Most arc welding robots are playback robots, and they perform excellently in replaying accurately what they are taught. Even so, the arc welding of structures is still susceptible to deviations in original weld lines, and bends or variations in groove condition arising from the degree of accuracy in assembly, tack welding or setting, or distortion during welding. Thus it is necessary to sense these variables and correct the robot's torch position or locus, or welding conditions during actual work time to allow unattended welding operation.

## 43.2 SHORT-ARC SENSOR

An arc sensor detects deviations from a weld line by sensing or measuring changes in welding current when a torch weaves in a groove. This sensing method has been available since the 1970s. However, since conventional arc

*Sensors and Control Systems in Arc Welding.* Edited by Hirokazu Nomura.
Published in 1994 by Chapman & Hall, London. ISBN 0 412 47490 5

sensors were developed to be used with thick plates, they were not sufficiently accurate with thin plates, which use a welding current within the range of the short-circuiting arc where arc conditions are not stable.

To solve this problem, the short arc sensor was introduced. The short arc sensor was developed by improving the output characteristic of welding power sources, by using a high-speed inverter, and by studying various data processing techniques.

### 43.2.1 Principle of the short-arc sensor

An arc sensor contains various processing techniques. Essentially, it determines and corrects any differences between the current values in the horizontal direction $(I_{R1} - I_{L1})$ by weaving the torch as shown in Figure 43.1. At the same time, it also corrects the vertical direction. However, making such a simple comparison has often caused periodic bead deviations due to the vibrating molten pool.

Therefore, to suppress the effect of the vibrating molten pool, we compared the difference between the maximum current value $I_{R1}$ and the minimum current value $I_{R2}$ $(I_{R1} - I_{R2})$ in one direction of the weaving path with that $(I_{L1} - I_{L2})$ in the other direction. In this way, we obtained favourable results with a weaving frequency of 1–5 Hz.

### 43.2.2 Improvement of the welding power source

As discussed above, arc sensors operate by detecting changes in welding current. Therefore, the welding current must communicate the groove condition correctly. However, the short arc state is largely influenced by a short-circuit current, and furthermore, conventional welding power sources that are suitable for the short arc state have a large internal inductance. Therefore, they require a long time for attenuation once a large current has been charged. Otherwise, they cannot communicate the groove condition.

This fact indicates that all we have to do is reduce the inductance and increase the constant voltage characteristics. Then the transient build-up of a short-circuiting current will become greater, and a stable short arc state will be unobtainable.

As a result, it is necessary to maintain the characteristics suitable for short-circuiting and for arcing. The short-circuit time for the short arc state is between 2 and 5 ms. Therefore we decided to use a high-speed inverter to control the desired output within this time period.

This has enabled us to control short-circuiting and arcing individually and to enhance the maintainability of a constant voltage characteristic during arcing. In addition, the response time for charging and discharging current has increased more than ten times over the conventional welding power source. Using this power source makes it possible to detect groove information accurately on the basis of changes in the current.

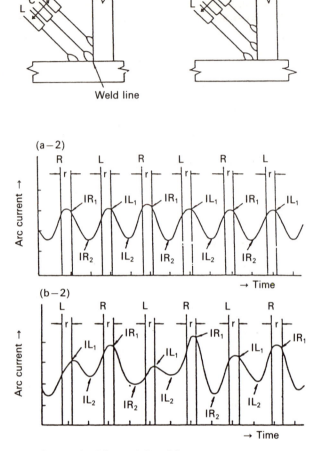

**Fig. 43.1** Arc sensing method for MAG welding.

Figure 43.2(b) shows the current difference $\Delta I$ plotted against the deviation distance $\Delta L$ with a conventional welding power source. Figure 43.2(a) plots the current difference $\Delta I$ against the deviation distance $\Delta L$ with the newly developed welding power source. Both the rate of current change and the correlation are greater in Figure 43.2(a).

### 43.2.3 Filtering

To obtain the signal required for tracking from the current signal, it is necessary first to pass it through the proper filter. The noise source in the current signal is mainly due to vibrations in the molten pool and short-circuiting. Vibrations in the molten pool relate to pool size, and can be specified by welding current and welding speed. The short arc sensor

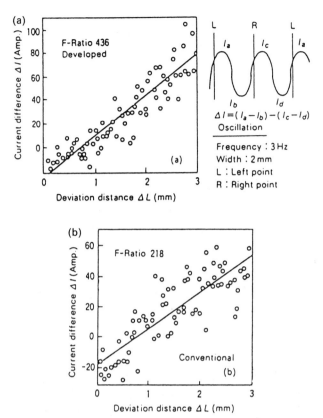

**Fig. 43.2** Relationship between deviation distance and difference in current: (a) newly developed power source; (b) conventional power source. Welding conditions: 140 A; 20 V; 40 cm/min.

determines the optimum digital filter by setting the data on the basis of weaving frequency, welding speed and wire supply speed.

### 43.2.4  Weaving frequency and its correction frequency

During the welding of thick plates, weaving is used to prevent incomplete fusion or other defects. The normal weaving frequency is 0.5–2 Hz. Arc sensors for thick plates can carry out one correction per weave. With a welding speed of 30 cm/min and a weaving of 2 Hz, a sufficient tracking performance has been obtained of one correction per 2.5 mm of weld line.

However, in the welding of thin plates, welding speed is higher, for example, for lap fillet welding of 2.4 mm, a welding speed of more than 1.2 m/min is required, and the arc sensors for thin plates require weaving of 4–5 Hz. Weaving of 0.5–2 Hz will be insufficient. The short arc sensor carries out weaving of 4–5 Hz and adopts a control method making two corrections per weave. The sensor is applicable with a maximum welding speed of

1.5 m/min for 2 mm lap fillet welding, thus assuring sufficient follow-up performance.

### 43.3   VISUAL SENSOR

When welding thin plates, it is necessary to compensate for such complications as workpiece shape errors, setting errors or thermal distortion. At the same time, it is also necessary to track small grooves very accurately at high speed to stabilize welding quality and improve productivity. A tracking system can be used for fillet welding (lap joint, T-joint) and butt joints (zero root gap, square groove, single V-groove) with plate thickness ranging from 0.5 to 4.5 mm. This system uses a visual sensor that is capable of tracking a weld line at a maximum welding speed of 3 m/min.

#### 43.3.1   System structure

The tracking system consists of a visual sensor, a welding robot which has a built-in sensor rolling mechanism, a sensor CPU and a controller into which the robot CPU is incorporated (Figure 43.3).

1. *Visual sensor*: Consists of a semiconductor laser stripe light projector and a CCD camera. It uses an interference filter to shield disturbance light away from the arc or spatter.
2. *Sensor rolling mechanism*: Rotates the sensor around the welding torch to allow it to track a weld line ahead of the torch.

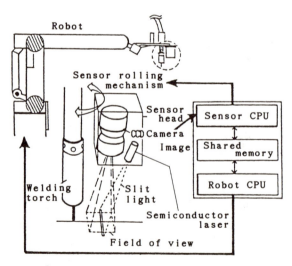

**Fig. 43.3** Structure of system.

3. *Controller*: A multi-CPU structure. Communication between CPUs is carried out through shared memory.
4. *Sensor CPU*: This function includes the detection of weld lines, the control of sensor rotations and the generation of tracking data.
5. *Robot CPU*: Operates a robot based on the tracking data issued by the sensor CPU through the shared memory.

### 43.3.2 Detecting weld lines

The visual sensor employs the optical profile method as the main principle of detection. The image processing flow presented in Figure 43.4 was adopted in order to solve the following problems:

1. butt joints with a zero root gap (the visual sensor cannot be used because no changes appear in the shape of the stripe);
2. the visual sensor cannot be applied to high-speed welding of more than 2 m/min.

The following section describes how butt weld lines were detected.

Figure 43.5 shows the intensity distribution of the stripe on a zero-root gap butt joint. Because of deterioration in the reflected intensity due to micro deformation at the cutting plane of the base metal, dark signals are

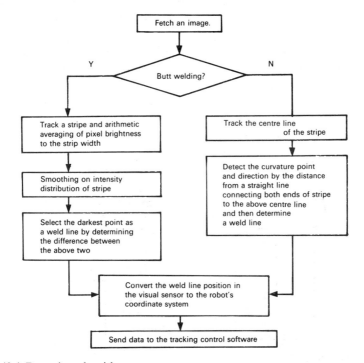

**Fig. 43.4** Detection algorithm.

generated on the weld line position. Since this detection processing is assured, the darkest point is considered to be the weld line after the removal of high-frequency noise in the intensity distribution of stripe due to speckles, and low-frequency noise due to the intensity distribution of the projector. For high-speed welding, the total field of view is not processed. Instead the window provided in the vicinity of the previously detected weld line position shortens the detection time.

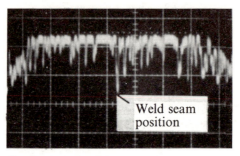

**Fig. 43.5** Intensity distribution of stripe.

### 43.3.3   Tracking control

We conceived a sequential tracking control method which takes weld line detection data directly as the robot's targeted travelling distance to improve the accuracy of the tracking control. We then provided a buffering function capable of asynchronizing the sensor CPU's detection timing and the robot CPU's control timing to maintain the degree of certainty (Figure 43.6).

The sensor CPU converts the weld line position, which is a two-dimensional image detected through image processing, to three-dimensional information by the triangulation method. It then computes the weld line position coordinates (DSP: detected seam position) in the robot's coordinate system.

The DSP is asynchronously transmitted to the robot CPU from the sensor CPU through the FIFO (first-in first-out) buffer of the shared memory. The sensor CPU generates the DSP and accumulates it sequentially in the FIFO buffer, while the robot CPU fetches the oldest DSP in the FIFO buffer when the wire tip reaches the previously removed DSP, and thereby controls the robot's new target position.

### 43.3.4   Application results

Figure 43.7 shows the welding results of a lap joint (plate thickness: 1 mm) to which the visual sensor was applied. The result of track welding the weld line of fillet welding (lap joint, T-joint) butt joints (root gap zero, square groove, single V-groove) with maximum welding speeds of 3 m/min resulted in favourable welding quality. Moreover, the tracking welding accuracy of

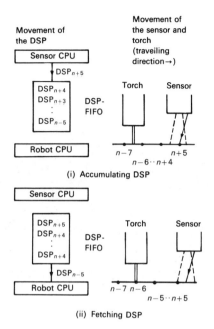

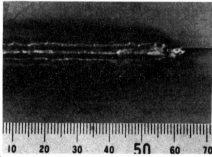

**Fig. 43.6** Torch position control method: (a) accumulating detected seam position; (b) fetching detected seam postition.

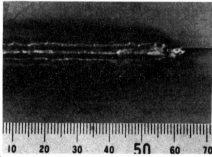

**Fig. 43.7** Weld bead appearance.

the straight line and the circular line was $\pm 0.3$ mm within the plane perpendicular to the torch and $\pm 0.5$ mm in the torch direction.

## 43.4  CONCLUSION

An arc sensor does not require a detection head bcause the arc itself can serve as a sensor, thereby avoiding the effects of the arc, spatter or fume. However, since it requires weaving, a practical welding speed range will be around 1.5 m/min. A visual sensor is applicable to a welding speed range up to a maximum of 3 m/min, at which the arc sensor cannot function.

CHAPTER 44

# An intelligent arc-welding robot with simultaneous control of penetration depth and bead height

*Yuji Sugitani, Yasuhiko Nishi and Toyoyuki Satoh*

NKK and Nippon Sanso KK have jointly developed a fully auto-mated intelligent arc welding robot for the fabrication of SUS 304 stainless steel cylindrical pressure vessels. In this system, the root gap is detected by image processing with a CCD camera, and then the welding parameters are controlled to keep penetration depth and bead height constant. This chapter describes the principle of simultaneous control of penetration depth and bead height and the features of the welding robot.

## 44.1 INTRODUCTION

In order to achieve full automation of arc welding, an intelligent arc weld-ing robot is required that can track the weld seam, detect the shape of the groove, and control the torch position and the welding para-meters.

An intelligent arc welding robot called 'Intelliarc' has been developed. This chapter describes Intelliarc's system and functions, including the method for simultaneously controlling the penetration depth and bead height with variable root gaps.

*Sensors and Control Systems in Arc Welding.* Edited by Hirokazu Nomura.
Published in 1994 by Chapman & Hall, London. ISBN 0 412 47490 5

## 44.2 ADVANCED FUNCTIONS OF INTELLIARC

### 44.2.1 High-speed rotating-arc welding

Intelliarc uses the high-speed rotating-arc welding method developed by NKK. The principle of the process is shown in Figure 44.1. The electrode rotates orbitally at a frequency of 20–60 Hz. For the work presented in this chapter, a frequency of 50 Hz and a 3 mm diameter orbit was used with 1.6 mm diameter flux-cored wire.

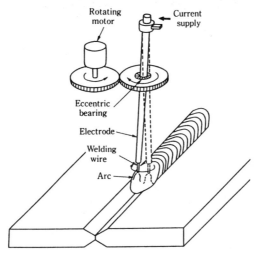

**Fig. 44.1** Principle of high-speed rotating arc welding.

The high-speed rotating-arc process ensures shallow and stable weld penetration and a smooth, level weld bead. This phenomenon appears to be caused by the dispersion of the arc force and heat with arc rotation.

### 44.2.2 Seam tracking and torch height control by arc sensor

The principle of seam tracking with arc sensor control is shown in Figure 44.2. The figure shows the basic waveform of arc voltage in relation to the rotating position of the arc. By dividing the wave form at $C_f$ (centre front of rotation) into left and right, and comparing the integrated values ($S_L$ and $S_R$) of these two regions, the aiming point of the torch can be detected.

This seam-tracking system has the following advantages compared with conventional arc sensors.

1. High-speed rotation enables remarkably quick control response.
2. High-speed rotation increases the detection sensitivity of torch deviation, and offers extremely accurate seam tracking.

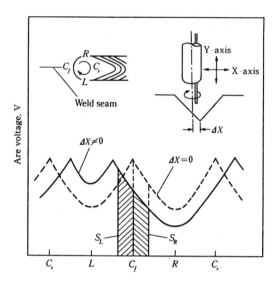

**Fig. 44.2** Principle of seam tracking by arc sensor.

Torch height control is achieved by comparing the average of the welding current waveform with a reference value.

### 44.2.3  Image-processing system

A diagram of the image-processing system is shown in Figure 44.3. The CCD (charge-coupled device) video camera is located about 100 mm in front of the welding torch, and detects information about the weld groove, including the root gap. Welding parameters are controlled by the detected values to keep the weld penetration and bead height constant with changes in root opening. Image processing is also used to set the initial torch height before welding and to detect the end of the weld seam.

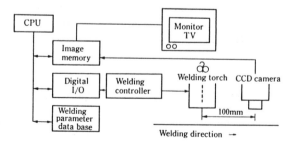

**Fig. 44.3** System diagram of image processing.

### (a)  *Method for measuring root gap*

The basic procedure of image processing for measuring the root gap is shown in Figure 44.4. The CCD camera is placed in front of the torch and provides an image of the groove as a view from above (Figure 44.4(a)). The transverse distribution of light intensity (brightness) along line A can be obtained as in Figure 44.4(b). The intensity is highest from the reflection of the arc on the bevel surface, while it is lowest in the root gap. This trend can be enhanced by using supple-mentary illumination. The root gap is obtained by processing these transverse brightness scans. The brightness is integrated to eliminate any noise component as shown in Figure 44.4(c). Subsequently, a smoothed distribution is obtained by differential treatment, which has peaks at positions where the intensity changes significantly (Figure 44.4(d)). The root gap is measured as the distance between the two inside peaks, while the groove width as the distance between the two outside peaks.

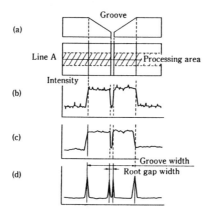

**Fig. 44.4** Basic procedure of image processing for measuring root gap width: (a) original image; (b) intensity distribution of brightness in transverse; (c) smoothing; (d) differential processing.

The root gap can be measured directly in this way for the first side welded. For the second side, however, only the groove width can be measured, since no root gap exists. In either case, a delay is introduced that corresponds to the distance that the camera is ahead of the torch.

### (b)  *Setting of the initial torch height*

In order to provide unmanned, remote-control welding the torch height can be set automatically before welding by image processing. The weld torch height is set by the focus of the CCD camera.

### (c)  Detecting the end of the seam

Intelliarc can detect the end of the weld seam and stop welding. The end of the workpiece is detected as the end of the weld preparation or the point where the groove is filled with weld metal.

### 44.2.4  Welding parameters control

Simultaneous control of penetration depth and bead height [2] is achieved by controlling four welding parameters; welding current, wire feeding rate, arc voltage and welding speed, based on changes in the root gap detected by image processing.

### (a)  Consideration of simultaneous control

A comparison of cross-sectional bead shapes with welding current is shown in Figure 44.5. With constant welding speed, an increase of welding current primarily causes bead width to be increased as shown in Figure 44.5(a). With welding speed control to keep bead height constant as shown in Figure 44.5(b), the depth of penetration changes dramatically owing to changes in welding current.

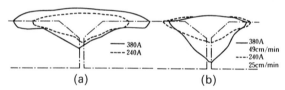

(a)                                  (b)

**Fig. 44.5** Influence of the welding current on the penetration depth: (a) constant welding speed; (b) with welding speed control to obtain constant bead height.

Thus both welding current and welding speed affect penetration depth. In the Intelliarc system, welding current is controlled to obtain constant penetration, while bead height is kept constant by modifying the welding speed to changes of the root gap.

### (b)  Welding speed control (bead height control)

The principle of using welding speed control to obtain constant bead height is shown in Figure 44.6. The change of deposition area $\Delta S(G)$ corresponding to a change in the root gap is calculated as shown by the shaded portion in Figure 44.6. Since $S_0$ is the optimum deposition area without a root gap, the welding speed $v$, which provides the necessary deposition area $S_0 + \Delta S(G)$, is determined by the root gap $G$ and the wire feeding rate $V_f$.

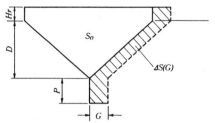

**Fig. 44.6** Principle of welding speed control to obtain constant bead height. $\Delta S(G) = (Hr + D + P) \times G$, $v = K V_f/(S_0 = \Delta S(G))$, where $v$ = welding speed (mm/s), $K$ = cross-sectional area of wire (mm²), $V_f$/wire feeding rate (mm/s).

### (c) Welding current control (penetration depth control)

The relationship between welding current and penetration depth for various root gaps is known from experience as shown in Figure 44.7. Intelliarc has a database of these relationships to determine the most suitable welding current for any root gap. For welding the second side, only welding speed control is used with a high constant current, since there is no risk of burn-through.

### (d) Wire feeding rate control (wire extension length control)

Wire feeding rate $V_f$ is controlled to maintain a constant electrode extension by using Lesnewich's [3] experimental equation:

$$V_f = AI + BLI^2 \qquad (44.1)$$

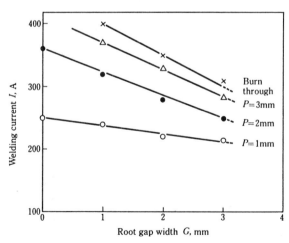

**Fig. 44.7** Relationship between root gap width and welding current to obtain constant penetration depth. Wire, 1.6 mm diameter stainless steel cored wire; gas, 100% $CO_2$; power source, constant potential characteristics; rotation of torch, 50 Hz, 3 mm diameter; $P$ = penetration depth.

where $I$ = welding current (A); $L$ = electrode extension (mm); and $A$ and $B$ are constants.

### (e)    Arc voltage control (arc length control)

The terminal voltage between the welding torch and the work piece, $E_t$, is controlled to maintain a constant arc length. $E_t$ is assumed to be the sum of the arc voltage $V_a$, and the voltage drop in electrode extension $V_L$, as illustrated in Figure 44.8.

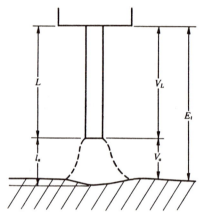

**Fig. 44.8** Definitions of arc welding parameters: $L$, wire extension length; $l_a$, arc length; $E_t$, terminal voltage between welding torch and workpiece; $V_L$, dropped voltage in wire extension; $V_a$, real arc voltage.

The voltage drop in electrode extension, $V_L$, can be obtained by Halmoy's [4] experimental equation:

$$V_L = aLI - \frac{bV_f}{I} \tag{44.2}$$

where $a$ and $b$ are constants.

Furthermore, the arc voltage can be estimated by Equation (44.3), which equates $V_a$ to the sum of the anode voltage drop, the cathode voltage drop and the arc column voltage drop. Equation (44.3) assumes that the sum of anode and cathode voltage drop is equal to the arc voltage just before short-circuiting, and that arc column voltage is equal to the product of a constant potential gradient of arc column, $X$, and arc length.

$$V_a = V_0(I) + Xl_a \tag{44.3}$$

where $V_0(I)$ is the sum of anode and cathode voltage drop as a function of welding current $I$(V); $X$ is the potential gradient of the arc column (V/mm); $l_a$ is arc length (mm).

### (*f*)  *Experimental results*

The simultaneous control of penetration depth and bead height was achieved with a double-Y groove, SUS 304 stainless steel test plate with a root gap varying from 0 to 3 mm. Parameters used were 15 mm for electrode extension $L$, 3 mm for arc length, $l_a$, 2 mm for penetration depth, and 1 mm for reinforcement height.

Figure 44.9 shows the variation of the root gap detected by image processing and the controlled welding parameters. The appearance and macrosections of the controlled weld bead are shown in Figure 44.10. These results indicate that both penetration depth and bead height are virtually constant.

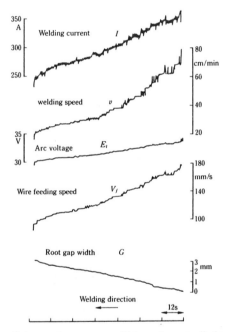

**Fig. 44.9** Variation of detected root gap width and controlled welding parameters.

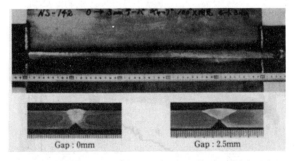

**Fig. 44.10** Appearance and macrosections of controlled weld bead.

## 44.3   PRACTICAL APPLICATION

The configuration of Intelliarc is shown in Figure 44.11. The welding head is mounted on a manipulator, and can be turned through 90° to enter the vessel through the 420 mm manhole. The robot can be applied to various sizes of vessel with a *y*-axis slider.

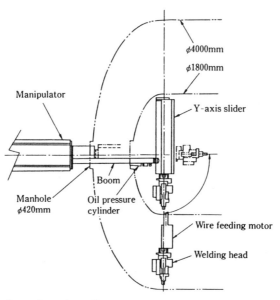

**Fig. 44.11** Configuration of Intelliarc.

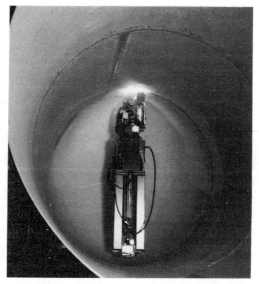

**Fig. 44.12** View of welding application.

Intelliarc has been successfully used to weld a liquefied gas storage tank at the Keihin Factory of Nippon Sanso KK. Figure 44.12 shows the final weld of the vessel's head.

## 44.4  CONCLUSION

Intelliarc has achieved the simultaneous control of penetration depth and bead height with continuously varying root gaps. The welding process is efficient and produces high-quality welds, while dramatically increasing operator safety.

## REFERENCES

1. Nomura, H. *et al.* (1986) Development of automatic fillet welding process with high speed rotating arc. IIW Document IIW.SG.212-655-86, IIW.XII-939-86.
2. Sugitani, Y. *et al.* (1987) Control of penetration depth and bead height in butt welding. Welding Arc Physics Committee of the Japan Welding Society, 87-657 (in Japanese).
3. Lesnewich, A. (1958) Control of melting rate and metal transfer in gas-shielded metal-arc welding. Part 1: Control of electrode melting rate. *Welding Journal*, August, 343-S.
4. Halmoy, E. (1979) Wire melting rate, droplet temperature, and effective anode melting potental. *Proc Int Conf on Arc Physics and Weld Pool Behaviour*, London.

# CHAPTER 45

# An NC welding robot for large structures

*Takaaki Ogasawara*

This chapter introduces the GT-5000 NC welding robot. The robot has been made available for practical use by Kobe Steel, and can be used to weld bridge beams and other large structures. This robot is capable of improving productivity to six times that of conventional manual welding and is highly acclaimed in Japan. Automation of welding large structures, which has been difficult using conventional teaching–playback robots or simple NC robots, was achieved by combining a number of sensor and data-processing technologies developed specifically for this robot.

## 45.1 INTRODUCTION

Changes in the working environment in recent years have caused chronic problems in labour shortage and the ageing of skilled workers. To resolve these problems, automation and robotization of welding work have been aggressively promoted. However, large structures such as bridge girders have hindered the progress of robotization for the following two reasons.

1. Each item is individually manufactured. There are almost no cases in which the same workpieces are repeatedly manufactured as they are in the automobile industry. The application of normal teaching–playback robots to such workpieces, therefore, takes a longer time because of the required teaching work, thereby making them inefficient.
2. Working accuracy cannot be maintained. Studies on NC robots have been done in which teaching was done numerically off-line, to improve the availability of the robots at the site. However, because the workpiece

*Sensors and Control Systems in Arc Welding*. Edited by Hirokazu Nomura.
Published in 1994 by Chapman & Hall, London. ISBN 0 412 47490 5

was deflected during setting or assembly, accurate workpieces could not be obtained. Errors were produced between the coordinates of the NC data and the actual workpiece and this resulted in the impracticability of NC robots.

To realize robotic welding of large structures, Kobe Steel has developed a high-performance NC robot (GT-5000) by adding various sensing functions to an NC robot and allowing the correction of the robot's position in accordance with the sensors' signals. This new NC robot solves the above-mentioned problems and has produced favourable application results.

## 45.2  STRUCTURE

The newly developed NC robot consists of an off-line teaching personal computer, which developes NC data, and an NC robot main body.

### 45.2.1   Off-line teaching personal computer

Large structures are often designed by the CAD method, and therefore one of two simplified teaching systems can be used: a high-order database system in which an operator extracts weld-line information from the CAD data and thereby develops a robot's operating program, or a system in which an operator inputs numerals into the drawing by interacting with the CRT display unit which employs computer graphics. Both of these systems are structured to output a robot's operating program automatically as well as the parameters required for sensing according to the welding material or the shape of the workpiece.

### 45.2.2   NC robot body

Figure 45.1 shows an external view of the GT-5000 NC robot.

The robot has a total of 12 controlling axes and can efficiently carry out horizontal fillet welding by using two torches.

The four main axes which move in accordance with the input NC data are:

1. a gantry travel carriage running along the track (x axis);
2. a traverse motion carriage moving along a gate beam (y axis);
3. an elevator moving up and down the head (z axis);
4. a revolving axis allowing the torch to face the welding direction ($\theta$ axis).

In addition, one set of three orthogonal axes holding torches is placed on either side to make up auxiliary axes that can absorb any error occurring between the NC data and the actual workpiece. One arc sensor oscillator is also provided on each side.

**Fig. 45.1** External view of the GT-5000 NC robot.

The control device is mounted on the carriage to allow for easy operation in accordance with the display on the CRT screen. A push/pull feeder has also been provided to allow the working rate to increase through the use of a bulk wire. Moreover, automatic cleaning of the torch nozzle allows the continuous welding distance to be longer.

### 45.3   PERFORMANCE

#### 45.3.1   Two-point shift

The NC data indicating the size and shape of a workpiece is shown in the workpiece coordinate system on the off-line teaching personal computer, which is independent of the robot (Figure 45.2). The robot, on the other hand, operates with the robot's coordinate system, thus requiring a conversion of coordinates. For this reason, in the NC data, a function which can automatically add two points for conversion at the diagonal line of a workpiece has been provided. An operator teaches the robot the two points displayed on the CRT by operating the robot and using the ITV.

To minimize errors, the two-point shift carries out coordinate transformation using the middle point (centre of gravity) between two points as centre of rotation.

During coordinate transformation, compatibility of the workpiece and its data to the robot's operating range is automatically checked.

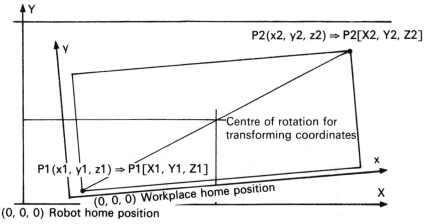

**Fig. 45.2** Two-point shift.

Furthermore, to avoid adverse effects of unevenness in the workpiece, a three-dimensional conversion is also available.

### 45.3.2 Wire ground sensor

The robot moves to a weld line in accordance with NC data. A wire ground function is provided to compensate for errors such as workpiece positioning errors (Figure 45.3). The wire ground function determines the correct weld line by detecting gaps between the wire and base metal when voltage is applied to the wire using auxiliary axes.

This function makes it possible to start the arc at the correct position.

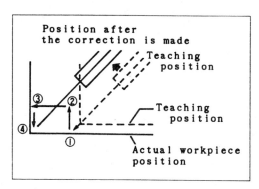

**Fig. 45.3** Wire ground sensor.

### 45.3.3 Edge sensor

Large structures require range definition at the welding start and end points. In range definition, finding the edge of the workpiece is important. Thus, the edge sensor detects the edge of the workpiece. To eliminate detection errors caused by unevenness in the workpiece edge (chips or scallops), an edge shape parameter is automatically added to the NC data. See Figures 45.4–45.6.

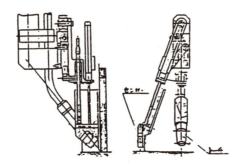

**Fig. 45.4** Edge sensor.

Scalloped          Flat          Snip

Sensor
detection
position

**Fig. 45.5** Shape of the edge.

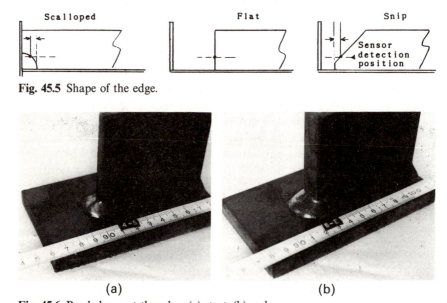

(a)                          (b)

**Fig. 45.6** Bead shape at the edge: (a) start; (b) end.

### 45.3.4   Arc sensor

An arc sensor is provided to detect changes in the current when the wire oscillates. This allows the torch to monitor and compensate for dislocations or distortions in the workpiece being welded.

Since changes in the current differ depending on the welding material or leg of the fillet weld, the arc sensor is structured so that the wire's melting characteristic or aimed shift distance is automatically added to the NC data as a sensing parameter.

In addition, internal inductance is minimized so that the welding current detects the size of the molten pool with the change in current. The arc sensor uses a power source for robots that provides a constant high voltage and allows stable sensing under a wide range of conditions. See Figure 45.7.

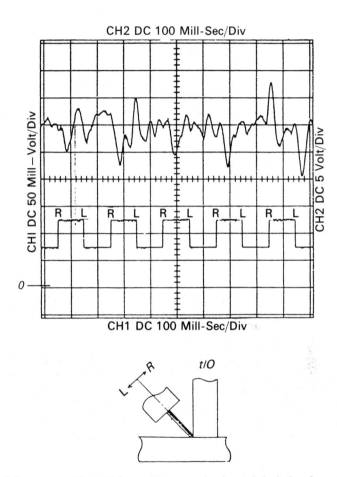

**Fig. 45.7** Arc sensor. Plate thickness 10 mm; no horizontal deviation from the target position.

### 45.3.5 Nozzle cleaning

One of the problems with MAG welding is that spatter generated during welding blocks the shielding nozzle resulting in insufficient shielding. Specifically, painted steel plates produce more spatter and often restrict the time for which robots can operate continuously.

The NC robot is equipped with an automatic nozzle cleaner and is structured to clean the nozzle automatically upon completing a welding job.

Because the amount of spatter is almost proportional to wire consumption, the pre-established wire consumption is used to determine when cleaning should begin. During the teaching work, nozzle cleaning is no longer a problem.

### 45.3.6 Application results

Results obtained by applying the NC robot to a bridge girder are shown in Figure 45.8 and Table 45.1. The arc-generation rate exceeded 65%. Productivity was as much as six times that when using semi-automatic welding.

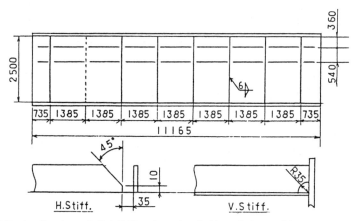

**Fig. 45.8** Application of GT-5000 robot to a bridge girder: weld geometry.

**Table 45.1** Application of GT-5000 robot to bridge girder: results

| | |
|---|---|
| Total weld distance | $37.28 \times 2 = 74.56$ m |
| Operating time | |
| Two-point shift | 80 s |
| Welding | 7946 s |
| Air cutting | 2250 s |
| Nozzle cleaning | 1440 s (180 s × 8 times) |
| Total operating time | 11716 s (3 hr 15 min 16 s) |
| Arc generation rate | 67.8% |

Note: Steel plate wash-primer-coated. Welding wire DW-300 1.4 mm diameter.
Welding condition:  Leg 4 180 A × 26 V × 30 cpm
Leg 6 210 A × 28 V × 30 cpm
Leg 8 210 A × 28 V × 20 cpm

## 45.4 CONCLUSION

Combining various sensor techniques with an NC robot makes it possible to automate/robotize the welding of large structures which was previously considered impractical.

Before using the NC robot in welding, the hardware aspects of robots were not the only consideration. It was also necessary to develop welding techniques that were suitable for robots and to find welding materials that could be applied to workpieces coated with paint or having steel plates.

In the future, improvement of data communication functions and simplification of the teaching work are desired so as to further the completeness of processing cells in CIM. Moreover, it is hoped that more functions will be made available so as to widen the range of applicability.

# CHAPTER 46

# Development of a fully automatic pipe welding system

*Akio Tejima*

We have succeeded in developing a fully automatic TIG welding system that enables continuous unmanned welding operations from the initial layer to the final finishing layer. This welding system is designed for continuous, multilayered welding of thick and large-diameter fixed pipes of nuclear power plants and large boiler plants where high-quality welding is demanded. In the tests conducted with this welding system, several hours of continuous unmanned welding corroborated that excellent beads are formed, good results are obtained in radiographic inspection and that quality welding is possible most reliably. This system incorporates a microcomputer for fully automatic controls, and features seam tracking, automated control of wire feed position, and self-checking for inter-pass temperature, cooling water temperature and wire reserve.

## 46.1  INTRODUCTION

For the thick and large-diameter fixed pipes of nuclear power plants and large boiler plants where high-quality welding is demanded, automation of welding has been positively promoted. But in conventional automatic welding, it has always been necessary for a skilled welder to monitor the arc and weld puddle, adjust the torch position and filler wire feeding position, and at the same time adjust the welding conditions (current, voltage, speed, etc.) in accordance with the situation, to secure high-quality welding. The authors have developed a fully automatic welding system called CAPTIG (Computer controlled All position Pipe TIG welding system) that can perform all the welding operations except the preparation works to be done by the welding operator.

*Sensors and Control Systems in Arc Welding.* Edited by Hirokazu Nomura.
Published in 1994 by Chapman & Hall, London. ISBN 0 412 47490 5

The welding operator only installs the track ring on the piping, clamps the welding carriage on it, positions the torch, sets the initial data, and pushes the start button (see Figure 46.1). Then the system can continuously perform unmanned welding operations from the penetration welding of the initial layer to the final finishing layer. This system, which has a continuous

**Fig. 46.1** Welding carriage and head.

inversion welding mechanism, has self-checking functions for interpass temperature, torch cooling water temperature and filler wire reserve as well as control functions such as microcomputer-aided integrated control, seam tracking and filler wire feeding position control, and can perform accurate, stable and high-quality welding operations.

This chapter will outline the system, centring on the control mechanism.

## 46.2  APPLICABLE PRODUCTS AND WELDING METHOD

The main specifications of the applicable pipes are:

Material: Carbon steel, stainless steel, chromium molybdenum steel
Pipe diameter: 200–800 mm
Wall thickness: 8–80 mm

The highly reliable TIG (tungsten inert gas) welding method was adopted, since this assures the best quality.

## 46.3  CONTROL MECHANISM FOR UNMANNED OPERATION

### 46.3.1  Flowchart of welding work

Figure 46.2 shows a flowchart of welding work. First, as preparation, the track ring is installed on the piping and the welding carriage is clamped on

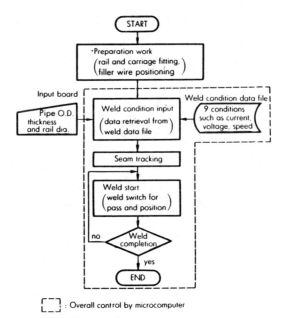

**Fig. 46.2** Flowchart of welding work.

it. After the torch and filler wire positions are adjusted, the outside diameter, wall thickness of the pipe, etc. are set as initial data on the control equipment.

After the above manual works are completed, all the remaining works are controlled by the microcomputer. The welding control unit already has the welding conditions classified into such items as material, joint geometry and welding position, stored as a database file in the microcomputer, and from them, suitable conditions for the initial layer to final layer are selected in accordance with the outside diameter and wall thickness of the pipe. The conditions for each layer can be programmed for each of the eight sections the pipe circumference has been divided into.

If the material, joint geometry or welding position is changed, the database is replaced with another one by inputting another file by means of a magnetic cassette tape.

### 46.3.2 Continuous inversion welding

If the welding carriage turns only in one direction when pipes are multilayer-welded continuously, the torch cable and head control cable will be wound up around the pipe. To prevent this, this system is so designed that the welding carriage turns in the other direction as it finishes one turn around the pipe.

### 46.3.3 Two-direction alternate filler wire feeding

This system performs continuous welding while repeating the inversion. Therefore, it is equipped with a wire feeder on both sides of the torch, and the wire feeders alternately operate in accordance with the turning direction and always feed the filler wire from the front in the welding direction.

### 46.3.4 Seam tracking

#### (a) Principle

This is a contact-type seam tracking method using the tungsten electrode itself as the sensor. Before welding starts, the electrode scans right and left inside the groove, detects its position when it contacts the groove face, seeks and stores the centre position of the groove and then outputs it when welding.

Figure 46.3 shows the principle of seam tracking and the calculation method for the centre position of the groove. The tungsten electrode is connected to the detecting power supply, which is separate from the welding power supply, and the torch is moved in the groove by means of the rack gear with a potentiometer at right angles to the welding direction. When it contacts the groove at $X_a$ in Figure 46.3 and the detecting circuit operates, the driving motor is immediately stopped, and the voltage $V_a$ of the potentiometer at that position is stored. Then the torch is moved in the reverse direction and the position indicating potential $V_b$ at the groove face $X_b$ is stored. Then the mean value of the position-indicating potentials at the right and left faces $V_c$ $(\frac{1}{2}(V_a + V_b))$ is calculated to obtain the centre position of the groove.

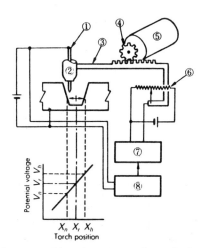

**Fig. 46.3** Principle of seam tracking: 1, tungsten electrode; 2, torch; 3, rack; 4, gear; 5, motor; 6, potentiometer; 7, average operator; 8, contact detector.

The above detecting operation is repeated at constant intervals and displacement of the entire weld line is stored. During seam tracking, these displacement data are sequentially output, and linear interpolation is made between the detection points by proportionally distributing the preceding and following values over the travel distance.

### (b)  Coping with welding distortion

Since this seam-tracking method detects the groove position prior to welding, coping with the distortion displacement caused during weld- ing is a problem. In particular, transverse shrinkage of the groove greatly affects the torch position. For this reason, the contraction of groove at the time of continuous welding is experimentally measured for each pass using the pipe material, diameter, wall thickness and groove geometry as parameters, and half of the contraction is programmed in the file of welding conditions as a shift amount of the torch position. Figure 46.4 shows one example of measurement results of groove con- traction.

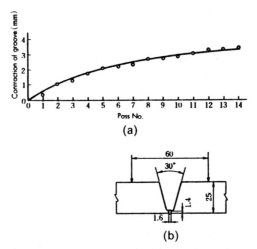

**Fig. 46.4** (a) Contraction of groove. (b) Groove size and shape: dimensions mm. Testpiece: carbon steel, outside diameter 267 mm.

### (c)  Results of seam-tracking test

Using a specimen of carbon steel plate with thickness of 20 mm with a groove (angle, 14°; bottom width, 6 mm; depth, 15 mm) bent at an inclina- tion of ±4°, then pass multilayer welding was performed at the flat position. The groove position detection was made only once prior to welding, and the tracking error at each layer was only about 0.3 mm at maximum. As Figure 46.5 shows, a good welding was achieved.

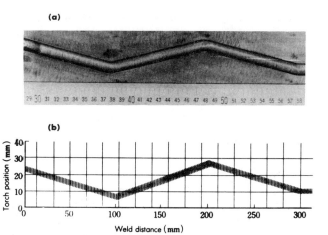

**Fig. 46.5** (a) Appearance of weld bead; (b) variation in torch position.

### 46.3.5   Automatic control of filler wire feeding position

In all-position welding of fixed pipes, the direction of the gravity working on the weld puddle changes through 360° and the shape of the weld puddle also changes accordingly. For this reason, the relative position between the filler wire and weld puddle changes momentarily. Therefore, in ordinary welding, the welding operator is always making adjustments to optimize the wire feeding position.

To automate this task, the system servo-controls the vertical position of the wire utilizing the small voltage occurring between the filler wire and base metal. If the voltage is measured between the wire and base metal when the wire is away from the weld puddle during welding a value of several volts, almost near to the arc voltage, will be obtained because of the space potential of the arc. But even when the wire is completely in contact with or buried in the weld puddle, a small voltage of tens of millivolts is still obtained, and this voltage changes depending on how the wire is fed to the weld puddle.

Figure 46.6 shows an example of measurement of the voltage between the wire and base metal when the filler wire is fed to the weld puddle. This is a record of voltage from the state where the wire is in contact with the interface between the molten puddle and solid metal (lower limit) to the state (upper limit) just before the wire leaves the weld puddle and becomes droplets after it moves upwards. The voltage between the filler wire and base metal shows about 30 mV at the lower limit and about 100 mV at the upper limit, and is almost proportional to the height of the wire-feeding position. If this voltage is fed back to the circuit to move the wire position vertically under the same principle as that of the AVC (automatic arc voltage control), the wire feeding height from the weld puddle can be arbitrarily controlled.

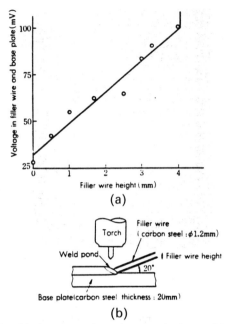

**Fig. 46.6** (a) Relationship between wire voltage and wire position. (b) Test method. Welding conditions: arc current, 200 A; arc voltage, 10 V; welding speed, 8 cm/min; wire feed rate, 100 cm/min.

## 46.4   WELDING EQUIPMENT

The welding equipment comprises track ring, welding carriage and head, control unit, welding power supply, and torch cooling water pump.

### 46.4.1   Track ring, welding carriage and head

The track ring is made light by using aluminium alloy and can be easily installed on the pipe because it can be divided into two parts by means of a hinge. The welding carriage is also made of aluminium and is equipped with two wire feeders on both sides of the moving part.

Since the straightness in feeding the filler wire is always kept within ±0.5 mm by means of the restraightening mechanism developed by IHI, it is not necessary to adjust the straightness during welding.

### 46.4.2   Welding control unit

This control unit (Figure 46.7) uses two 8-bit microprocessors (i-8085). By dividing the control role into two parts, one for controlling the entire welding work and the other for controlling the seam tracking, the operation is speeded up and the software development is made easier. Figure 46.8

**Fig. 46.7** Control unit.

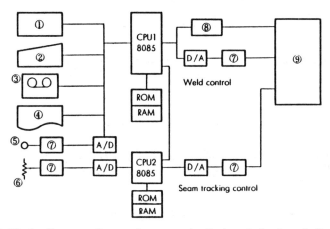

**Fig. 46.8** Block diagram of control system: 1, display; 2, keyboard; 3, magnetic cassette; 4, printer; 5, temperature sensor; 6, potentiometer; 7, isolation amplifier; 8, photocoupler; 9, controller.

shows the block diagram of the microcomputer system of this equipment. Except for the display and keyboard installed near the microprocessors, all the interfaces with the sensor, sequence control and actuator control are isolated with a photocoupler and isolation amplifier to prevent the invasion of noise.

## 46.5 WELDING RESULTS

The system was used to perform the unmanned welding of horizontal fixed pipes. Table 46.1 shows the welding conditions. The weld obtained was good in appearance for both penetration bead and surface bead and the radiographic test revealed no defects.

**Table 46.1** Welding conditions

| Item | Number of passes | | |
|---|---|---|---|
| | *1* | *2* | *3–14* |
| Arc current (A) | 160 | 180 | 210–230 |
| Arc voltage (V) | 9.5 | 9.8 | 10.0–10.8 |
| Weld speed (cm/min) | 8.5 | 8.0 | 7.0–9.5 |
| Wire feed rate (cm/min) | 65 | 80 | 85–120 |
| Oscillation width (mm) | — | 3.5 | 4.0–12.0 |
| Oscillation speed (cm/min) | — | 60 | 100–120 |
| Groove shape and size (dimensions in mm) | | | |

Note: Specimen material, carbon steel
Outside diameter, 267 mm
Process: one layer, one pass
(total 14 layers)

## 46.6   CONCLUSION

The all-position welding of thick and large-diameter fixed pipes of nuclear powers plants and large boiler plants demands high technology, but the conventional automatic welding machine requires monitoring and adjustment by a skilled welding operator and does not differ so much from manual welding in that the welding quality depends on the skill of the operator. The development of this system, which realizes full automation of this type of welding, is a big step towards the realization of a consistent factory automation-oriented pipe production line as well as towards the high-level stabilization of welding quality. The features of this system are:

1. integrated control by means of microcomputer;
2. continuous inversion welding;
3. two-direction alternate filler wire feeding;
4. seam tracking;
5. automatic control of filler wire feeding position;
6. self-checking function.

# Index

Page numbers appearing in **bold** refer to figures and page numbers appearing in *italic* refer to tables.

ACC (automatic current control) 5
Acoustic sensors, *see* Sound sensing
Adaptive control 5, 14–16
  with arc phenomena sensors 15–16
  MRAC (model reference adaptive control) 58–62
  on-line optimization 62–4
  with optical sensors 16, 160–7
  with seam tracking control 16
  with temperature sensors 14–15
AE (acoustic emission) 41
AI (artificial intelligence)
  narrow-gap GMA welding 185–8
  robots 69–70
Arc current sensing, *see* Weld current sensing
ARC-EYE visual arc sensor 132–3
Arc length control 396
Arc sensors
  limitations (from questionaire) *84*, 84–5
  *see also* Arc short sensing; Optical arc image sensors; Optical linear sensors; Optical point sensors; Weld current sensing
Arc short sensing, back bead control 279–84
Arc sound sensing, *see* Sound sensing
Area (image) optical sensors, *see* Optical arc image sensors
AVC (automatic voltage control) sensing, *see* Weld current sensing
Ayrton's welding current equation 231

Back bead control 31, 279–84
  from molten pool width 47–8

Backing plate voltage, bead width control 373–6
Bead height control 261–2
  by weld current sensor 271–8, 394–5
  facing bead control by filler wire feed 377–9
Bead width control, SAW welding
  arc intensity sensing 370–3
  copper backing voltage sensing 373–6
  laser sensing 192–4
BHK narrow-gap welding process 182–90

CAD (computer-aided design) 75
CAM (computer-aided manufacture) 75
Cameras, *see* CCD cameras; ITV (industrial TV) cameras
CAPTIG (computer controlled all position pipe TIG welding system) 408–16
  application 409, 415–16
  continuous inversion welding 410
  control flowchart 409–10
  equipment 414–5
  filler wire feed control 413–14
  seam tracking 411–12
  two-direction alternate filler wire feed 411
  welding distortion 412
CCD cameras 33–4
  with high speed rotating arc welding 392–4
  molten poll observation 35–6
  pulse TIG MIG MAG welding 36–7
  *see also* ITV cameras
$CO_2$ arc welding
  arc light spectrum 92–3

417

$CO_2$ arc welding—*contd*
  arc sound monitoring 310–15
  deposition control 29–30
  fuzzy control for 137–45
  one sided with back bead control 31
  process 258, 264–5, 282–4
  thin steel sheets 111
Contact probe sensors 4
  contouring 301–2
  for seam tracking 8, 18–20
    torch aligned 19–20
    using memory playback 19
    using teaching playback 19
    torch angle control 21
    with welding of LNG corrugated
      membranes 300–7
Control systems
  classification 7–17
  definition 7
  deposition control 44–5
  future expectations 86–7, 87
  state equation development 45–6
  *see also* Adaptive control; Fuzzy
    control; Monitoring; Robots,
    off-line programming/teaching;
    Seam tracking
Control theory
  adaptive control 58–64
  Laplace transforms 54, 55
  sampling 53
  weld pool width simulation 58
  z transforms 54–6
Corrugated plate welding 21

Deposition control, with weld current
  sensing 28–30, 276–7
Digital filters, applied to arc-current
  sensing 223
DSP (detected seam position) 388

Edge detection, Gt-5000 NC welding
  robot 403–4
Electrode contact sensors (touch
  sensors) 4–5
  applications 289–91
  construction 293
  fillet sensor 295, 296
  groove detection 297–9
  groove gap width sensing 288
  groove sensor 295
  on GT-5000 robot 403
  intermittent sensing 287
  multiple sensing 296–7
  multipoint sensing 287, 296–7

one-point sensor 294, 296
  performance 202–3
  principle 199–202, 285–6, 294
  for seam tracking 8, 20–1
  Shin Meiwa's touch sensor 293
  single-axis sensor 295
  starting point sensing 236–7, 286
  stick type sensing 288
  three-dimensional sensing 286
  two-point sensor 294
  weld length sensing 287
Electromagnetic sensors
  capacitance types 39
  eddy current types 39–40
  principle 6
  seam tracking 10–11
  ultra-heat-resistance 323–32
    application 330–1
    basic elements **324**
    construction 325
    overall function 329–30
    requirements 323–4
    temperature compensation 325–8

FCB-SAW, *see* SAW
Fuzzy control
  $CO_2$ short-arc welding 137–45
  control rules 49–51, 64, 66, 67
  fuzzy inference method 64–7, 67
  molten pool control 67–8
  neural networks 149–52
  seam tracking 49–51, 147–53, 238–46
  with welding current sensing 238–46

Gap control (for welds), ARC-EYE
  135
Gas sealed thermometers 5
Generamometer (oscillating mirror)
  system 104–10
GMA welding
  automatic narrow-gap welding 182–90
  narrow groove detection and control
    38–9
  one-sided 279–84
  torch position control 210–14, 299
Groove gap sensing 288
Groove tracking
  by CCD camera and image
    processing 392–4
  by rotary mirror displacement
    sensor 161–4
  differential distance laser sensing 89,
    90–1

laser sensing by optical beam
scanning, method 88–9, 90
*see also* Seam tracking; Vibrating reed
sensor
Groove width sensing 22, 45, 260–1,
297–9
molten pool optical sensing 36–7
*see also* Vibrating reed sensor
GT-5000 welding robot, *see* Robots,
NC (numerical control)
GTA welding 210, 299

Halmoy's wire melt rate equation 230
HAZ (heat affected zone) 24
Hitachi M6060 welding robot 112
Hitachi MOS image device (HE97211)
113

IC card welding control system 350–9
applications **354**, 358–9
education training unit 357–8
general configuration 351
specification *352*
Image processing
multilayer welding 128, 162–5
narrow-gap GMA welding 183–5
narrow gap MIG welding 178–9
short-circuiting arc welding 37–9
stripe images 114–15
Infrared semiconductor sensors 22, **23**,
24
Intelliarc system 390–9
Intermittent sensor 287
ITV (industrial TV) cameras 33–4
with adaptive control 161
infrared vidicon 176–9
with narrow gap MIG welding
168–78
with remote-controlled
circumferential TIG welding
361–2
three-dimensional position
measurement 97–8, 128–31
weld line detection 386–8
*see also* CCD cameras; Optical arc
image sensors

JC (joint capacity), reverse bead
control factor 192–3

Lap joint welding 206–7
Laplace transforms 54, 55
Laser diodes, Ga-Al-As 113

Laser sensing
laser stripe optical sensor
automatic threshold adjustment
99–100
light sector weld position
detection method 96–7
multilayer welding 95–103, 127–
31
optical filter 99–100
robot control 111–17, 118–26
seam tracking 115–16, 125
sensor specification *102*
three-dimension position
measurement 97–8
optical beam scanning
differential distance method 89,
90–1
groove tracking control 88–94,
161–2
multilayer welding 168–75
SAW welding with flux copper
backing 195
oscillating mirror (generamometer)
system
performance *107*, 107–10
principle 105–6
system construction 106
Linear optical sensors, *see* Laser
sensing; Optical linear sensors
Line Master for Daihen robots 225–7
Liquid sealed thermometers 5
LNG corrugated membranes
automatic welding 300–7
welding parameters *306*

MAG arc welding
adaptive control with visual sensing
160–7
arc sensors 25
back bead control 31
by GT-5000 robot 405
molten pool observation 36
pulsed welding control by weaving
width 272–4
seam tracking and deposition
control 29–30, 44–5, 266–78
thin steel sheets 111
Magnetic control, high speed TIG
welding 154–9
MIG welding
automatic narrow-gap welding 176–
81
molten pool observation 36
molten pool width control 56–8

Molten pools
  control with fuzzy inference 67–8,
    138–45
  control with MRAC 58–62
  on-line optimization, model based
    62–4
  sensing 45–7
  state equation 46
  width control 56–8
    with fuzzy inference 67–8
Monitoring welds 16–17
MRAC (model reference adaptive
    control) 58–62
Multi-electrode automatic fillet welding
    equipment 253–4
Multilayer welding
  with adaptive control 162–7
  with arc current sensing 263–4
  laser scanning 95–103, 127–31, 168–
    75
Multi-point sensor 287

Narrow gap welding, with high speed
    rotating arc 252–3
NC (numerical control), *see* Robots,
    NC
Neural networks, with fuzzy variables
    149–52
NGW welding process 182–90
Nozzle cleaning 405

Off-line programming/teaching, *see*
    Robots, off-line programming/
    teaching
Optical arc image sensors 7
  adaptive weld control with 16, 160–
    1, 162–7
  applications 34–9
  ARC-EYE
    construction 132–3
    principle 133–6
    seam tracking function *136*
  fuzzy control, $CO_2$ welding 137–46
  with glass fibre optics for magnetic
    control of arc 154–9
  narrow gap GMA welding 182–90
  narrow gap MIG welding 176–81
  seam tracking with 13
  *see also* CCD cameras; ITV; Optical
    linear sensors; Optical point
    sensors
Optical CCD camera groove detection
    392–4
Optical filters 100

Optical linear sensors 7
  characteristics 32–3
  seam tracking 11
  *see also* Laser sensing
Optical point sensors 7
  back bead control, one sided
    welding 30–1
  back sensing for back bead width
    control 370–3
  characteristics 32–3
  seam tracking 11
Optical sensors 6–7
  applications *34*, 34–9
  characteristics 32–4
  *see also* Laser sensing; Optical arc
    image sensors; Optical linear
    sensors; Optical point sensors;
    PSD (position-sensitive
    detector); Rotary mirror laser
    displacement
Optical weld line image sensing, seam
    tracking with fuzzy logic 147–53
Oscillating mirror (generamometer)
    system 104–10

PAW (plasma arc welding), back bead
    width control 374
Penetration depth control 395
Perche cooling module 113
Pipe welding
  automatic with IC card system 358–
    9
  remote-controlled circumferential
    TIG 360–8
  *see also* CAPTIG
Point optical sensors, *see* Optical
    point sensors
Popov's hyperstabilization theory 59
Pressure vessels, narrow-gap MIG
    welding 176–81
Process control, *see* Control
PSD (position-sensitive detector) 32,
    120
Pyrometers
  colour temperature 22
  monochromatic temperature 22

Quartz rod, for spectral analysis 35
Questionaire analysis 76–85
  future trends 86–7
  sensor advantages/disadvantages **82**,
    83–5, *84*
  sensor applications 78–9, **79**
  sensor industrial usage **78**, 78

sensor operation classification 78–9, **80**

sensor usage growth **80**, **81**, 82

Radiation thermometers 5
Remote-controlled circumferential TIG welding 360–8
Robots
   architecture 74
   articulated arc welding 254–6
   collision avoidance 338–41
   control
      seam tracking 111–17, 125, 236–7
      start point finding 122, 236
   control of penetration depth and bead height 390–8
   Hitachi M6060 112
   intelligent 69–70
   layers of control 69–70
   Line Master for Daihen robots 225–7
   motion generation 336–40, 344–6
   motion generation, torch 335–6
   multilayer welding 127–31
   NC (numerical control) GT-5000
      applications 406–7
      arc sensor 404–5
      edge sensor 403–4
      nozzle cleaning 405–6
      structure 401–2
      two-point shift 402–3
      wire ground sensor 403
   off-line programming/teaching 73–5, 333–41
      applications 347–9
      width CAD 333, 344–5, 401
      GT-5000 robot 401
      hardware for 343
      operating procedure 343–4
   safety 349
   sensors for 69–75
      algorithms 72–3
      interface 72–3
      requirements 70–1
   teaching method 73
   torch position control 21, 210–14
   with visual seam tracking sensor 111–17, 118–26
Rotary mirror laser displacement sensor 161–2
Rotating-arc welding, *see* Weld current sensing

Safety, robots 349

Sampling 53
SAW (submerged arc welding), one-sided
   arc light intensity back bead control
      control system development 372–3
      principle 30–1, 370–2
   face bead height control
      control system development 378–9
      principle 377–8
   laser diode detector system 191–8
   voltage detection back bead width control
      control system development 376
      principle 373–5
Seam tracking
   with adaptive control 16
   CAPTIG pipe welding system 408
   contact probes for 8, 18–20
   with deposition control 44–5
   with electrode contact sensor 8, 20–1
   electromagnetic sensors for 10–11
   fuzzy control based 48–52, 147–53, 238–46
   industrial applications 78
   with ITV weld line detection 386–9
   with laser stripe and visual sensor 111–15
   with optical area sensor 13, 38–9, *136*, 136, 147–53
   with optical linear sensor 11
      robot control 125
   with optical point sensor 11
   with scanning laser optical sensor 168–72
   with ultrasonic sound sensor 13–14
   with weld current sensing 8–10, 25–9, 209–13, 267–8
      with high speed rotating arc 248–52, 391–2
   *see also* Groove tracking
Sensors
   applications
      growth in use **80**, **81**, 82
      industrial usage by industries **78**, 78
      usage by applications 78–9, **79**
   basic requirements 2
   classification 3–7
      by operational method 78–9, **80**
   definition 2
   detection objectives 3
   future expectations 86–7, *87*
   with robots
      connections 72–3

Sensors—*contd*
  offline programming 73–5
  requirements 69–71
  *see also* Contact probe sensors;
    Electrode contact sensors;
    Electromagnetic sensors; Laser
    sensing; Optical sensors;
    Pyrometers; Seam tracking;
    Sound sensing; Temperature
    sensors; Weld current sensing
Shin Meiwa's touch sensor 293
Short-arc sensor, with weld current
  sensing 382–6
Short-circuiting arc welding, image
  processing for 37
Simulation, arc sensing using fuzzy
  control 243–6
Sound sensing 7, 40–1
  AE (acoustic emission) 41
  process monitoring
    $CO_2$ welding results 310–15
    principles 309
    TIG welding results 310–15
  with remote-controlled
    circumferential TIG welding
    361–3
Spatter 122
Speed control, with contact sensors
  303–5
Speed control of welding, adaptive
  control 165

TCB-SAW, *see* SAW
Teaching robots, *see* Robots, off-line
  programming/teaching
Temperature sensors 5, 22–4
  adaptive control with 14–15
  infrared semiconductor 22, **23**
  *see also* Gas sealed thermometers;
    Liquid sealed thermometers;
    Radiation thermometers;
    Thermister thermometers;
    Thermocouples
Thermister thermometers 5
Thermocouples 5
Through-the-arc sensors, *see* Weld
  current sensing
TIG (tungsten inert gas) welding
  arc light spectrum 92–3
  arc sound monitoring 310–15
  CAPTIG 408–16
  control by weaving width and area
    274–8
  high speed by magnetic control
    154–9

iron spectrum observance 35
LNG corrugated membranes 300–7
molten pool observation 36–7
narrow groove hot wire 24
optimization by molten pool
  observation 62–4
remote controlled circumferential
  system 360–8
seam tracking and deposition
  control 28–9, 266–78
Torch angle control 21, 210–14
Torch oscillation, *see* Weld current
  sensing
Touch sensor, *see* Electrode contact
  sensors
T.V. cameras, *see* CCD cameras; ITV
  cameras

Variability of welded objects 1–2
V groove tracking, *see* Groove
  tracking
Vibrating reed sensor
  excitation 319
  principle 317
  structure 317, **318**
  tapping sound detection 319–21
Visual arc sensors, *see* Optical arc
  image sensors
Visual laser sensing, *see* Laser sensing

Weld current sensing
  alternative terms for 5
  for deposition control 28–30, 404–5
  torch oscillation/weaving
    application 222–4
    bead height control 261–2
    CAPTIG seam tracking system
      411
    dynamic analysis 216–22
    equations for 25, 230–2
    with fuzzy control 238–46
    groove width tracking 260–1
    power source improvement 383–4
    principle 5, 25–7, 203–6, 229–30,
      257–60, 382–3
    for seam tracking 27–30, 209–13,
      259–60, 267–9
    sensing performance 233–6
    with short-arc sensing 383
    weaving control 267–71
    weaving frequency 385–6
  torch rotating at high speed 25–30,
    250–6, 391–2, 394–9
  *see also* Arc short sensing

Weld line tracking, *see* Seam tracking
Weld pools, *see* Molten pools
Wire earth/ground/touch sensors, *see*
    Electrode contact sensors

YASNAC-ERC 132–3

Z transforms 54–6, 57–8